OXFORD CHEMISTRY PRIMERS

Physical Chemistry Editor	Founding/Organic Editor	Inorganic Chemistry Editor	Chemical Engineering Editor
RICHARD G. COMPTON	STEPHEN G. DAVIES	JOHN EVANS	LYNN F. GLADDEN
University of Oxford	University of Oxford	University of Southampton	University of Cambridge

Polymers

David J. Walton

Coventry University

J. Phillip Lorimer

Coventry University

OXFORD

UNIVERSITY PRESS

This book has been printed digitally and produced in a standard specification
in order to ensure its continuing availability

OXFORD
UNIVERSITY PRESS

Great Clarendon Street, Oxford OX2 6DP

Oxford University Press is a department of the University of Oxford.
It furthers the University's objective of excellence in research, scholarship,
and education by publishing worldwide in

Oxford New York

Auckland Cape Town Dar es Salaam Hong Kong Karachi
Kuala Lumpur Madrid Melbourne Mexico City Nairobi
New Delhi Shanghai Taipei Toronto
With offices in
Argentina Austria Brazil Chile Czech Republic France Greece
Guatemala Hungary Italy Japan South Korea Poland Portugal
Singapore Switzerland Thailand Turkey Ukraine Vietnam

Oxford is a registered trade mark of Oxford University Press
in the UK and in certain other countries

Published in the United States
by Inc., New York

ISBN 0-19-850389-X

Antony Rowe Ltd., Eastbourne

Series Editor's Foreword

Oxford Chemistry Primers are designed to provide clear and concise introductions to a wide range of topics that may be encountered by chemistry students as they progress from the freshman stage through to graduation. The Physical Chemistry series contains books easily recognised as relating to established fundamental core material that all chemists will need to know, as well as books reflecting new directions and research trends in the subject, thereby anticipating (and perhaps encouraging) the evolution of modern undergraduate courses.

In this Physical Chemistry Primer Phil Lorimer and David Walton present a clearly written and stimulating introductory account of Polymer Chemistry considering both fundamental aspects and applications. The book explains in simple terms the basic ideas of the subject and outlines the importance of polymeric materials in the modern world; essential knowledge for all chemists. This Primer will be of interest to all students of chemistry (and their mentors).

Richard G. Compton
Physical and Theoretical Chemistry Laboratory
University of Oxford

Preface

Polymers are the archetypal modern materials, and they have transformed the world in which we live. This Primer covers the span of polymer types and their uses, starting from natural polymers, with a review of polymer history, followed by discussion of the statistical nature of polymer chains, from which originates their physical properties, and in particular the ease of processing that allows the production of complex shapes, which is a great benefit in technology. The main polymer types are discussed in detail, with emphasis upon practical commercial considerations and description of real industrial processes. The behaviour of polymer networks is explained, and finally the modern importance of functional polymers is discussed, in which a property in addition to great molecular size is exploited, for example electrical conductivity. It is intended that the reader will gain an overview of key aspects of this important class of compounds.

DJW
JPL

Contents

1 General principles and historical aspects

1.1 Introduction

This primer is intended as an introduction to the methods of producing synthetic polymers and to show how the structures of these materials give the properties which are important in daily life. The organization of the book is designed to provide an overview, graduated from simple systems through to more complicated modern polymer technology. Thus Chapters 1 and 2 provide general principles, while Chapters 3 and 4 treat two main types of polymers and their practical industrial applications respectively. Chapter 5 addresses hybrid systems while Chapter 6 addresses more sophisticated modern functional polymers that have unusual properties.

1.2 Principles and terminology

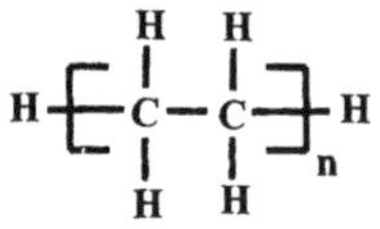

Polymers are formed by linking large numbers of small molecules together. The term polymer derives from the Greek 'poly' meaning 'many' and 'mer' meaning 'units'. Perhaps the simplest of all macromolecules is poly(ethene) (Fig. 1.1) which can be viewed as simply an extension of the covalently bound micromolecule ethane. Here n represents the number of $(CH_2–CH_2)$ units making up the polymer chain.

Fig. 1.1 Poly(ethene).

The value of n is termed the **degree of polymerization** $\overline{X_n}$. Since $\overline{X_n}$ could easily be a number as high as 10 000, then for a linear polymer with a repeat unit of relative molecular mass 28 (e.g. ethene), this would give a polymer of relative molar mass (RMM) 280 000. In other words

Molar mass of the chain = Molar mass of the repeat unit times $\overline{X_n}$

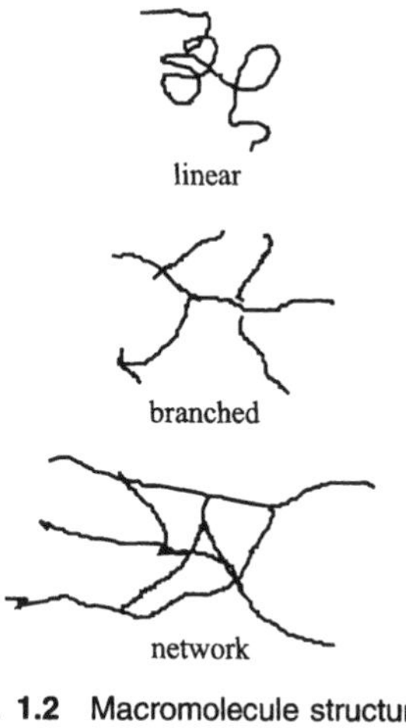

Other terms encountered are 'oligomers' meaning 'several units' (i.e. a very small polymer), trimer (three units, $n = 3$), dimer (two units, $n = 2$), and monomer, which is the basic unit from which a polymer is constructed. Macromolecules are therefore simple extensions of micromolecules in which chains of atoms are held together by covalent bonds. The chemical bonds within the chain are very strong and directional along the chains and, although linear chains are most commonly and easily produced, by changing the chemistry involved chains can be linked together to produce both branched and networked structures (Fig. 1.2).

Polymers are now ubiquitous in daily life. They have taken over from previous structural materials such as wood and stone, or from fibres used in the manufacture of clothing such as silk, cotton and wool (Table 1.1). Their increased usage and success has been based on economic factors such as the discovery of oil and the fact that as natural materials become scarcer they

linear

branched

network

Fig. 1.2 Macromolecule structure.

Table 1.1   A comparison of natural materials *versus* synthetic polymers

Natural	Synthetic substitute
Stone	Concrete
Wood	Polystyrene, bakelite
Silk	Nylon
Cotton	Terylene
Leather	Styrene–butadiene rubber, polyurethane
Natural rubber	Polyisoprene, neoprene
Glue (from bones)	Polyvinyl acetate, epoxy resins
Shellac (varnish)	Polyurethane, alkyd (polyester) resin

become relatively more expensive. For example, cast iron gutters and buckets have been replaced by poly(propene) and polyvinyl chloride, woodworking glue (originally made from the extract of bones) by polyvinyl acetate, and natural rubber and gutta percha by artificial polyisoprene. Table 1.1 gives a flavour of cases where some naturally occurring polymers have been replaced by synthetic (artificial) analogues.

Not only have polymers found a wide use in structural and textile materials, polymer substitutes have also found a wide application in medicine: for example, the use of polyethene hip joints, perspex in artificial corneas, nylon in arteries, and silicone rubber in artificial hearts.

Synthetic polymers have had a sizeable impact in the field of fibres, plastics and rubbers (elastomers), but they also exist in other categories and Table 1.2 lists typical examples and the industries in which they are used.

While the majority of polymers listed in Table 1.2 are synthetic polymers, there are natural polymers such as proteins, natural rubber, and cellulose each of which could be fitted into a similar category.

Of the various polymer categories in Table 1.2, fibres, plastics and elastomers (rubbers) are probably the most important technologically. Fibres are thought of as being very strong with good pliability, resistant to abrasion

Table 1.2   Synthetic polymer categories

Nature	Example	Industry
Fibres	Courtelle	Textile
	Terylene	
Elastomers	Natural rubber	Car tyres
	Synthetic rubber	Contraception
Plastics	Polytetrafluoroethene (PTFE)	Metal spinning, domestic utensils
	Polyvinyl chloride (PVC)	Electrical insulation
	Celluloid	Films
	Cellulose	Cosmetics, textiles, packaging
	Cellophane	Packaging
	Perspex	Optical
Resins	Araldite	DIY, glues
	Bakelite	Electrical insulation
	Isopon	DIY (fibre glass)
Surface-	Polyurethane	Paint
Coats	Alykd polyester	Paint
Foams	Polyurethane (flexible)	Packing, car parts
	Polystyrene (rigid)	Building, food processing

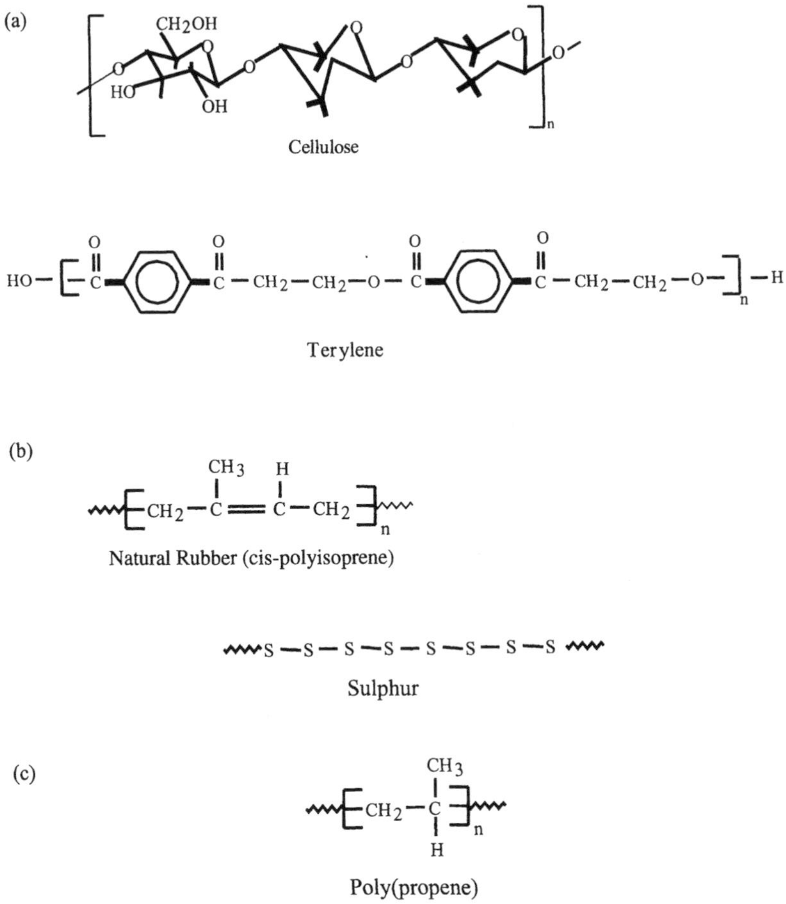

Fig. 1.3 Structures of (a) fibres, (b) elastomers, and (c) plastics.

and able to hold their shape, while an elastomer should be flexible, able to return to its original dimensions, and must not deform of its own accord. Plastics are seen as intermediate between the above two, being tough, having reasonable strength, some flexibility, and ability to retain their shape. A clue to the difference in the properties of fibres, elastomers, and plastics lies in the difference in their structures (Fig. 1.3)

For example, whereas the fibres cellulose and terylene contain ring structures to assist rigidity, the elastomer poly(isoprene) contains double bonds leading to the possibility of linking between chains (see later) to provide rigidity via a network; plastics contain mainly covalent chains with no functionality available for interchain linkage.

It is clear from Tables 1.1 and 1.2 that, while the term 'plastic' is often used to describe polymers, the words 'polymer' and 'plastic' are not synonymous and strictly the term 'plastic' should only be used to describe the stress–strain behaviour rather than refer to the chemistry. For example plastic sulphur, as it is known, formed by quenching molten sulphur in water, was the first true plastic material before the discovery of natural rubber. The subsequent discovery of a host of other materials which exhibit plastic deformation has revolutionized materials science and is the basis of the great usefulness of polymers.

Other advantages of polymeric materials over conventional materials are that they are light and easy to transport, easy to repair and highly resistant to corrosion, solvent action and moisture. However, they possess disadvantages in that they can have poor mechanical strength, a short lifespan and poor resistance to temperature.

Another feature of polymer physical properties concerns the time-dependence of mechanical behaviour. This phenomenon, known as visco-elasticity, is shown for example by the toy 'silly putty', which, if stretched rapidly, snaps, but if pulled slowly elongates elastically (see Chapter 4). The rich span of mechanical and physical properties shown by polymers is their great contribution to science.

Polymers also possess another property, namely processability. Previously only metals and glass were truly melt-processable, while the other natural materials such as stone, wood, and leather required mechanical cutting and then gluing or stitching into final shape. In fact, the widespread use of electricity benefited greatly from having plastic insulation extrudable onto wires, instead of the laborious wrapping of each wire with oil-soaked paper or similar material. This is just one example of the advantages of polymer processability in daily life.

1.3 General principles of polymerization

Step-growth polymerization

Polymers produced by this mechanism usually involve the typical condensation reactions of organic chemistry where a small molecule, e.g. H_2O or HCl, is expelled as the link is built. For example:

$$\text{e.g.} \quad \underset{\text{acid}}{CH_3C}\!\!\overset{\displaystyle O}{\overset{\displaystyle \|}{—}}\!OH + \underset{\text{alcohol}}{HOCH_3} \longrightarrow \underset{\text{ester}}{CH_3C}\!\!\overset{\displaystyle O}{\overset{\displaystyle \|}{—}}\!O — CH_3 + H_2O$$

$$\text{or} \quad R—CO—Cl + H—O—R_1 \longrightarrow R—CO—OR_1 + HCl$$

$$\text{Also} \quad \underset{\text{amine}}{R—NH—H} + \underset{\text{acid}}{HO—OCR_1} \longrightarrow \underset{\text{amide}}{R—NH—OCR_1} + H_2O$$

It requires little imagination to see that any such reaction, on paper at least, provides a route to polymer chains provided the starting materials are **bifunctional**. For example, taking X = OH (alcohol) and Y = COOH (carboxylic acid) gives an ester which is capable of further reaction with another unit (Figs 1.4 and 1.5).

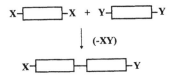

Fig. 1.4 Condensation of monomers

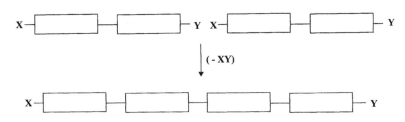

Fig. 1.5 Condensation of dimers

Choosing the synthesis of two very common step-growth polymers, the polyester terylene and the polyamide nylon as examples gives:

$$HO-\overset{\overset{O}{\|}}{C}-C_6H_4-\overset{\overset{O}{\|}}{C}-OH + H-O-CH_2{-}CH_2-O-H \longrightarrow HO-\left[\overset{\overset{O}{\|}}{C}-C_6H_4-\overset{\overset{O}{\|}}{C}-O-CH_2{-}CH_2-O\right]-H$$

TERephthalic acid	eth YLENE glycol	Repeat unit
(Benzene 1,3 dicarboxylic acid)	(1,2 dihydroxyethane)	Polyethylene terephthalate (PET)

$$H-NH-(CH_2)_6-NH-H \qquad + \qquad HO-\overset{\overset{O}{\|}}{C}-(CH_2)_4-\overset{\overset{O}{\|}}{C}-OH$$

Hexamethylene diamine	Adipic acid
1,6 diaminohexane	hexane 1,6 dioic acid

$$H-\left[-NH-(CH_2)_6-NH-\overset{\overset{O}{\|}}{C}-(CH_2)_4-\overset{\overset{O}{\|}}{C}-\right]-OH$$

6,6 nylon

(In the synthesis of polyamide nylon, (6,6) refers to the number of carbon atoms in each of the monomers).

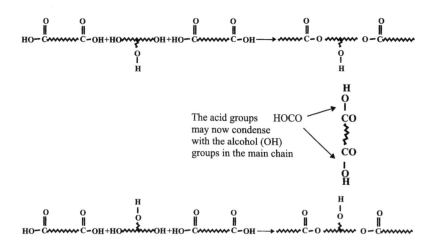

Fig. 1.6   An example of a network polymer.

The use of multifunctional monomers can produce network polymers. Figure 1.6 gives an example of a polymer produced from a trifunctional alcohol and a difunctional carboxylic acid.

The production of terylene shown earlier is by the condensation reaction between a diacid monomer and a diol monomer. Provided that the monomer contains both a hydroxyl group (X) and an acid group (Y) condensation will take place. However, there is a subtlety of structure best seen by drawing out several repeat units:

Here all the ester linking units are in the same directional sense:

However, we could produce an isomer of the same polymer by reacting the following:

If we draw the repeat units again, the linking units now reverse directional sense each time. This can affect crystal packing, hydrogen bonding, and polarity and hence polymer physical properties. (see nylons chapter 4, and kevlar^{TM} chapter 6)

Another way of influencing the physical properties (e.g. crystallinity, melting point, etc.) within a single class of polymer is to alter the proportions of the polar linking groups within the chain. Using nylons as example of a series of related compounds the effect of altering the length of the hydrocarbon units between the polar NH/CO links can be seen. For example 6,10 nylon can be prepared from the amine $H-HN-(CH_2)_6-NHH$ and the carboxylic acid $HO-CO-(CH_2)_8-CO-OH$ (or acid chloride $Cl-CO-(CH_2)_8-CO-Cl$); 8,6 nylon can be prepared by reacting $H-HN-(CH_2)_8-NHH$ and $HO-CO-(CH_2)_4-CO-OH$, and 6 nylon can be prepared by self-condensation of the aminocarboxylic acid $NH_2-(CH_2)_5-COOH$. The spacing and regularity between the NH and CO groups in nylon 6,6 or nylon 6,10 will obviously be different. Therefore for identical RMM more CONH groups would be expected to be present in 6,6 than 6,10 nylon and, if these groups participate in interchain hydrogen bonding, as they do, more hydrogen bonding is expected in 6,6 than 6,10. The consequence is that more thermal energy is required to separate chains composed of 6,6 than 6,10 and thus 6,6 has a higher melting point. The length of the hydrocarbon $(CH_2)_n$ sections also affects flexibility and mechanical properties as well as hydrophobicity and water repellency in the polymer.

While the ultimate aim in any synthesis is to obtain a particular RMM (and hence particular physical and mechanical properties), other processes may intercede. For example, the acid unit may decarboxylate to some degree under the conditions of polymerization (e.g. high temperature). This would give $Y = H$ as the end-group which, in the case of polyester production, would be unable to react with the alcohol, thereby terminating the polymerization reaction prematurely. Secondly, it may be advantageous to stop the reaction when a particular RMM has been achieved (i.e. a given value of n). For the polyester, addition of a monovalent alcohol $R'OH$ or monovalent carboxylic acid $R''COOH$ could be used as a chain stopper to seal one end of the polymer chain.

Chain polymerization

This is the second common method of synthesizing polymers. The word chain refers to the mechanism of polymerization, which in many cases is technically also an addition reaction, hence this term is sometimes used. The most common type of addition polymer is based on ethene $CH_2=CH_2$, in which the monomer contains at least one double (pi) bond which, on being activated (see Chapter 3), opens up to produce two single sigma bonds and the homopolymer poly(ethene) (Fig. 1.7). Note the resultant polymer backbone is joined together by carbon-carbon bonds, unlike the condensation polymer systems where there are oxygen, nitrogen or other heteroatoms in the chain.

If two monomers, e.g. ethene $CH_2=CH_2$ and propene $CH_2=CH(CH_3)$, are reacted together then there is the possibility of producing a copolymer. Figure 1.8 is an example of the copolymer produced by reacting ethene and propene. The actual structure of the copolymer will depend on the reaction condition and is explained in more detail in Chapter 3.

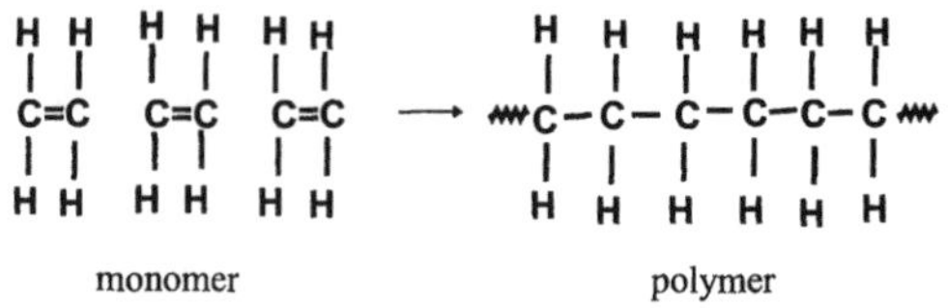

$$
\begin{array}{ccc}
\overset{H}{\underset{H}{\overset{|}{\underset{|}{C}}}}=\overset{H}{\underset{H}{\overset{|}{\underset{|}{C}}}} & \overset{H}{\underset{H}{\overset{|}{\underset{|}{C}}}}=\overset{H}{\underset{H}{\overset{|}{\underset{|}{C}}}} & \overset{H}{\underset{H}{\overset{|}{\underset{|}{C}}}}=\overset{H}{\underset{H}{\overset{|}{\underset{|}{C}}}}
\end{array}
\longrightarrow
\text{www}\overset{H}{\underset{H}{\overset{|}{\underset{|}{C}}}}-\overset{H}{\underset{H}{\overset{|}{\underset{|}{C}}}}-\overset{H}{\underset{H}{\overset{|}{\underset{|}{C}}}}-\overset{H}{\underset{H}{\overset{|}{\underset{|}{C}}}}-\overset{H}{\underset{H}{\overset{|}{\underset{|}{C}}}}-\overset{H}{\underset{H}{\overset{|}{\underset{|}{C}}}}\text{www}
$$

monomer   polymer

Fig. 1.7 Addition polymer.

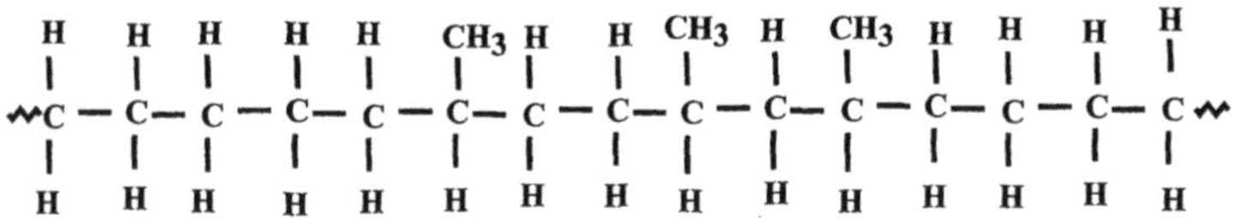

$$
\text{wwC}-\overset{H}{\underset{H}{\overset{|}{\underset{|}{C}}}}-\overset{H}{\underset{H}{\overset{|}{\underset{|}{C}}}}-\overset{H}{\underset{H}{\overset{|}{\underset{|}{C}}}}-\overset{H}{\underset{H}{\overset{|}{\underset{|}{C}}}}-\overset{CH_3}{\underset{H}{\overset{|}{\underset{|}{C}}}}-\overset{H}{\underset{H}{\overset{|}{\underset{|}{C}}}}-\overset{H}{\underset{H}{\overset{|}{\underset{|}{C}}}}-\overset{CH_3}{\underset{H}{\overset{|}{\underset{|}{C}}}}-\overset{H}{\underset{H}{\overset{|}{\underset{|}{C}}}}-\overset{CH_3}{\underset{H}{\overset{|}{\underset{|}{C}}}}-\overset{H}{\underset{H}{\overset{|}{\underset{|}{C}}}}-\overset{H}{\underset{H}{\overset{|}{\underset{|}{C}}}}-\overset{H}{\underset{H}{\overset{|}{\underset{|}{C}}}}-\overset{H}{\underset{H}{\overset{|}{\underset{|}{C}}}}\text{w}
$$

Fig. 1.8 Copolymer.

A general representation for most simple monomers would be:

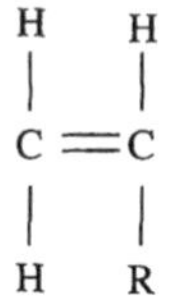

$$
\overset{H}{\underset{H}{\overset{|}{\underset{|}{C}}}}=\overset{H}{\underset{R}{\overset{|}{\underset{|}{C}}}}
$$

Before the introduction of systematic nomenclature, the group $CH_2{=}CH$ was known as the **vinyl** group, thus making monomers with $R = Cl$ (chloride) or $OCOCH_3$ (acetate) vinyl chloride and vinyl acetate respectively. Three typical monomers with $R = H$, Cl and $OCOCH_3$ are given in Table 1.3 together with their trivial and systematic name and the acronymic name for the polymer.

Table 1.3 Nomenclature of monomers and polymers

R	Monomer (trivial)	Monomer (systematic)	Polymer (acronym)
H	Ethylene	ethene	Polyethylene (PE)
C1	Vinyl chloride	chloroethene	Polyvinyl chloride (PVC)
$OCOCH_3$	Vinyl acetate	ethenyl ethanoate	Polyvinyl acetate (PVAc)

1.4   The statistical nature of polymer chains

An important distinction between macromolecules and small molecules can be seen when the homologous series of the alkanes (ethane, propane, butane, pentane, hexane) is considered, e.g.

CH_3CH_3	$CH_3CH_2CH_3$	$CH_3CH_2CH_2CH_3$	$CH_3CH_2CH_2CH_2CH_3$	$CH_3CH_2CH_2CH_2CH_2CH_3$
ethane	propane	butane	pentane	hexane

We could redraw ethane, butane, and hexane as shown in Fig. 1.9 and could imagine that if $n = 1$ we have ethane, for $n = 2$ we have butane and $n = 3$ we have hexane.

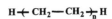
$$H{-}(\!CH_2{-}CH_2\!)_n H$$

Fig. 1.9 Alkane structure.

Small molecules such as alkanes have distinct identities and are always treated as separate compounds. For example, ethane is a gas, hexane is a liquid and octadecane (H-$(CH_2$-$CH_2)_9$-H) is a solid at room temperature. However, as the value of n increases the differences become less clear. For example, if $n = 50$ we have a relatively short-chain poly(ethene), although as prepared the end-groups are unlikely to be H, while if $n = 1000$ we have a high polymer of poly(ethene) with a molar mass of 28 000. These will differ in their physical and mechanical properties, but will otherwise be chemically identical, being effectively long chains of -CH_2-CH_2- units. The same principle applies to the long-chain isomeric polyamides. As the chain becomes long, they may be considered chemically identical and differences in their properties originate from differences in chain length.

In a real polymer sample consisting of many chains there is not usually a single value of n for all chains in the sample. Regardless of how much the polymer chemist may wish to produce molecules of one size, e.g. $n = 1000$, in practice a spread of RMMs will arise, as shown in Fig. 1.10, with some molecules having an n value larger than 1000, and some molecules having a smaller n value. In this regard polymer science lags behind nature, which is able to produce protein polymers of exact dimensions. All ethene molecules would simply appear as a point in Fig. 1.10 at RMM = 28.

If we consider the (self) condensation of hydroxyhexanoic acid, HO-RCO_2-H where $R = (CH_2)_n$ and where $n = 5$, the concept of an RMM distribution is easy to visualize. The first reaction to occur is between the end-groups on a pair of monomers to eliminate water (Fig. 1.11) and produce the dimer.

Since the end-groups are still active the next reaction of the dimer H-$(ORCO)_2$-OH can either be reaction with another monomer (M_1) or reaction with a dimer (M_2) to yield a trimer (M_3) and a tetramer (M_4) respectively, e.g.

$$H\text{-}(ORCO)_2\text{-}OH + H\text{-}ORCO\text{-}OH \rightarrow H\text{-}(ORCO)_3\text{-}OH + H_2O$$

$$H\text{-}(ORCO)_2\text{-}OH + H\text{-}(ORCO)_2\text{-}OH \rightarrow H\text{-}(ORCO)_4\text{-}OH + H_2O$$

In general, therefore, any reaction can be represented as:

$$H\text{-}(ORCO)_i\text{-}OH + H\text{-}(ORCO)_j\text{-}OH \rightarrow H\text{-}(ORCO)_{i+j}\text{-}OH + H_2O$$

Obviously from such a random set of reactions (Fig. 1.12) a distribution of RMM will be observed. In the case of chain addition polymerization, there will be a need to activate the monomer (ethene say, CH_2=CH_2) before polymerization can occur. The detailed mechanisms will be dealt with in Chapter 3 and the discussion here will be restricted to chain free radical polymerization.

H-ORCO-OH   +   H-ORCO-OH

$\downarrow$ (-H_2O)

H-ORCO -ORCO-OH   i.e. H-$(ORCO)_2$-OH

↑   ↑

able to react   able to react
with OH of acid   with H of hydroxy

Fig. 1.11 Self condensation of a hydroxy acid.

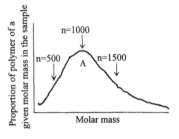

Fig. 1.10 Typical polymer RMM distribution.

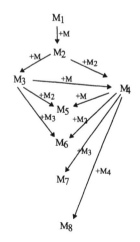

Fig. 1.12 Monomer, dimer and oligomer reactions.

A free radical is an uncharged species with an unpaired electron. It is highly reactive and undergoes reactions to extract an electron from another substrate so that it can produce a full complement of electrons within the newly created molecule. If a representative radical is termed R then reaction with a monomer such as ethene may be written as:

$$R\bullet \; + \; \underset{\substack{|\\H}}{\overset{\substack{H\\|}}{C}}=\underset{\substack{|\\H}}{\overset{\substack{H\\|}}{C} \; \longrightarrow \; R-\underset{\substack{|\\H}}{\overset{\substack{H\\|}}{C}}-\underset{\substack{|\\H}}{\overset{\substack{H\\|}}{C}\bullet$$

This new species is still a free radical with an unpaired electron, and it can continue to attack another ethene monomer (i.e. propagating or polymerizing) without the loss of activity. The radical can only lose (or terminate) its activity by reacting with another radical to form a covalent bond (Fig. 1.13).

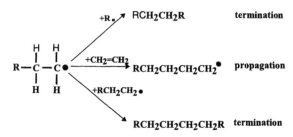

Fig. 1.13   Reactions of radicals.

A consequence of the reaction scheme in Fig. 1.13 is that the termination reaction of the dimer radical ($RCH_2CH_2CH_2CH_2\bullet$) with the activator radical ($R\bullet$) gives exactly the same molecule ($RCH_2CH_2CH_2CH_2R$) as the termination reaction between two monomer radicals ($RCH_2CH_2\bullet$). It is the random nature of these initiation and termination reactions which produces a mixture of chain lengths. As an example of the diversity of possible chain lengths that could be present in a hypothetical addition polymerization, Fig. 1.14 charts the progress of a radical R ($OH\bullet$ from the homolytic cleavage of H_2O_2 say) coming into contact with four ethene monomers. For simplicity the monomer unit will be represented by M, the dimer unit by M_2, trimer unit by M_3 etc. Also included in the figure are the RMM of each species.

Thus both step-growth and chain mechanisms give a spread of chain lengths and RMMs in the resultant polymer sample, and the question is which molar mass should be quoted — the lowest, the highest, or should a range be given? Although it is possible to obtain the spread of RMMs in a particular sample, in fact this is what was done previously to construct the RMM distribution (Fig. 1.10), it is a very tedious and time-consuming exercise. Therefore what is actually quoted is an average value. However, to do this some property of the polymer sample must be measured, against which to set the average. The easiest method is one which will count the number of molecules of a particular RMM. For example, with a total of 300 molecules in

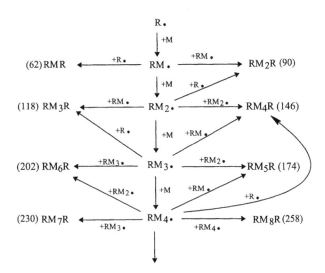

Fig. 1.14 Chain addition of four monomers. (Mass values refer to R as •OH and M as $CH_2 = CH_2$)

the sample, of which 150 molecules have an RMM of 100 000 and 150 molecules have an RMM of 300 000, then intuitively the average would be calculated at 200 000. Likewise, with a total of 300 molecules, of which 100 had an RMM of 100 000, 100 had an RMM of 200 000, and 100 had an RMM of 300 000, then again intuitively the average would be 200 000. Mathematically the value could be deduced from eqn 1.1:

$$\overline{M_n} = \frac{\sum N_i \times M_i}{\sum N_i} \qquad (1.1)$$

Here $\overline{M_n}$ is the **number average molar mass** and N_i is the number of chains (molecules) with an RMM of M_i. Using the three chain example we obtain:

$$\overline{M_n} = \frac{\sum (100 \times 100\,000) + (100 \times 200\,000) + (100 \times 300\,000)}{\sum (100 + 100 + 100)} = 200\,000$$

Any method which counts the number of molecules in solution is termed a colligative method. Particular methods include the lowering of vapour pressure, depression of freezing point, elevation of boiling point and osmotic pressure (see Chapter 2). These methods work well for small-molecule compounds, provided there is no dissociation, aggregation, or impurities present which will alter the total number of particles and hence the observed RMM value. Colligative methods are also applicable to polymer samples but, because of their great size, the actual number of macromolecules present in any polymer sample is not as high as for a sample of a micromolecular compound so that a smaller effect is observed for freezing point depression, boiling point elevation, osmotic pressure etc. The fact that this small effect is more difficult to measure helped fuel the controversy that existed from the nineteenth century through to the early part of the twentieth century over whether macromolecules actually existed.

Obviously there are many ways of taking averages. For people it could be the colour of their eyes, the colour of their hair, their height, or the *number* in a group, or the weight, or *mass*, of a group. Thus, it is also possible to obtain average molar mass values based on the size of a molecule (i.e. mass). The techniques commonly used to determine the mass average are chromatography (especially high performance gel chromatography, HPGPC), light-scattering (turbidity), ultracentrifugation or sedimentation (see Chapter 2).

In a similar definition to $\overline{M}_n$ (eqn 1.1) we can define the **mass average molar mass**, $\overline{M}_w$ (eqn 1.2):

$$\overline{M}_w = \frac{\sum W_i \times M_i}{\sum W_i} \tag{1.2}$$

where W_i is the mass (weight) of chains (molecules) with RMM of M_i. Substitution of the various masses of material of a given chain length (i.e. molar mass M_i) into eqn 1.2 will allow a deduction of $\overline{M}_w$. By making use of the fact that both the mass (W_i) and the number of molecules (Ni) are related to the number of moles (N_i) of any species present through the molar mass M_i and the Avogadro number N_A, it is possible to rewrite eqn 1.2 in terms of the numbers of molecules being counted rather than the mass of the molecules:

$$\overline{M}_w = \frac{\sum W_i \times M_i}{\sum W_i} = \frac{\sum N_i \times M_i^2}{\sum N_i \times M_i} \tag{1.3}$$

Also by analogy eqn 1.1 can be written as eqn 1.4:

$$\overline{M}_n = \frac{\sum N_i \times M_i}{\sum N_i} = \frac{\sum W_i}{\sum W_i/M_i} \tag{1.4}$$

Perhaps the best way to illustrate the difference between $\overline{M}_n$ and $\overline{M}_w$ in these equations is to insert a few numbers into the above equations. For the following calculations it is assumed that a polymer sample only contains two types of macromolecule, those with an RMM of 10 000 and those with an RMM of 100 000.

1. *Assuming **equal numbers** (N) of molecules*
   Using eqns 1.1 and 1.3 gives:

$$\overline{M}_n = \frac{\sum N_i \times M_i}{\sum N_i} = \frac{(N \times 10\,000) + (N \times 100\,000)}{(N + N)} = \frac{110\,000\,N}{2\,N} = 55\,000$$

$$\overline{M}_w = \frac{\sum N_i \times Mi^2}{\sum N_i \times M_i} = \frac{(N \times 10\,000 \times 10\,000) + (N \times 100\,000 \times 100\,000)}{(N \times 10\,000) + (N \times 100\,000)}$$

$$\approx 92\,000$$

2. *Assuming **equal masses** (W) of molecules*
   Using eqns 1.4 and 1.2 gives:

$$\overline{M}_n = \frac{\sum W_i}{\sum W_i/M_i} = \frac{(W + W)}{(W/10\,000 + W/100\,000)} \approx 18\,000$$

$$\overline{M_w} = \frac{\sum W_i \times M_i}{\sum W_i} = \frac{(W \times 10\,000) + (W \times 100\,000)}{(W + W)} = 55\,000$$

What is apparent from these calculations is that no matter whether equal numbers or equal masses are chosen the value of $\overline{M_w}$ is always greater than $\overline{M_n}$ (Fig. 1.15).

The ratio $\overline{M_w}/\overline{M_n}$ is termed the **heterogeneity index (HI)** or **polydispersity index**. A high $\overline{M_w}$ is usually required if the article is to have high ultimate strength (see later), whereas a low $\overline{M_n}$ is required for ease of flow of the polymer in the manufacture of the articles by injection moulding or extrusion. Therefore there needs to be a subtle balance of $\overline{M_w}/\overline{M_n}$ since any decrease in the ratio leads to an increase in impact strength, tensile strength, toughness, softening point and resistance to environmental stress cracking but with some loss in processing characteristics.

The $\overline{M_w}/\overline{M_n}$ ratio is also used as a measure of polydispersity of a system. Intuitively it would seem that a lower value of dispersity, i.e. a narrow spread of RMMs in a sample, is more desirable for control of physical and mechanical properties. Figure 1.16 represents a polydisperse system in which three molar mass ranges are identified. Clearly an aim of the producer of this material might be to try to improve the polymerization reaction to give a more uniform RMM distribution.

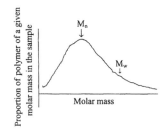

Fig. 1.15 RMM distribution showing M_n and M_w.

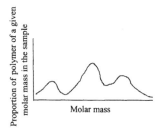

Fig. 1.16 Representation of a polydisperse system.

1.5   History of polymers

While the polymers discussed so far have been fully artificial macromolecules produced by step-growth or chain reaction of small-molecule species, there are a number of naturally occurring polymers. It was as a result of early attempts to modify these natural materials that the important understanding of polymer behaviour arose.

Natural polymers

Proteins

Proteins are based on α-amino acids (Fig. 1.17) and are regarded as the archetypal natural polymers from which many biological materials derive. They are complex molecules, with the individual repeat units containing some 24 different R groups (Table 1.4). They have an enormous range of molar mass and each protein has exactly the correct amino acid residue in the correct place along the chain. At present it is beyond the ability of artificial systems to produce polymers with the different units all in exactly the correct position.

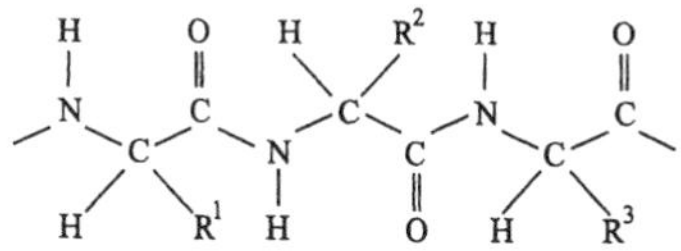

Wool is an example of a protein (keratin) which is also present in hair and is used as tough fibres to make clothing. Another tough natural fibre is cellulose, which is a chain of polyether-linked sugar rings (see Fig. 1.3a). It is found in a pure state in cotton which has long been used as fibres for clothing. As cotton, cellulose is a tough, insoluble and infusible fibre, which can only be cut, woven or mechanically manipulated. The structure has substantial opportunities for hydrogen bonding between rings and chains, via hydroxyl groups and also via the ether oxygens, and its high melting point (beyond decomposition temperature) and its insolubility originate because forces in the

Fig. 1.17 Representative structure of a segment of protein. Structures of R groups are given in Table 1.4.

Table 1.4   Naturally-occuring amino-acids

Structure of R in $H_2N^1-\overset{\overset{\displaystyle H}{|}}{C}-\overset{\overset{\displaystyle O}{\|}}{C}-OOH$
$\quad\quad\quad\quad\quad\quad\quad\quad\quad\underset{R}{|}$

Nonpolar R-Group	—H	Glycine
	—CH$_3$	Alanine

	—CH—CH$_3$
	  $\mid$
	  CH$_3$
		Valine

	—CH$_2$—CH—CH$_3$
	   $\mid$
	   CH$_3$
		Leucine

	—CH—CH$_2$—CH$_3$
	  $\mid$
	  CH$_3$
		Isoleucine

	—CH$_2$—CH$_2$—S—CH$_3$	Methionine
	—CH$_2$—◯	Phenylalanine
	—CH$_2$—◯—OH	Tyrosine

	—CH$_2$—(indole)
	    N
	    $\mid$
	    H
		Tryptophan

	$-O-\overset{\overset{\displaystyle O}{\|}}{C}-CH-CH_2$
	     $\underset{H_2N}{|}\quad\underset{CH}{|}$
	        CH$_2$
	complete structure
		Proline

Polar R-Group	—CH$_2$—OH	Serine

	—CH—OH
	  $\mid$
	  CH$_3$
		Threonine

	—CH$_2$—SH	Cysteine
	$-CH_2-\overset{\overset{\displaystyle O}{\|}}{C}-NH_2$	Asparagine
	$-CH_2-CH_2-\overset{\overset{\displaystyle O}{\|}}{C}-NH_2$	Glutamine
Acidic R-Group	$-CH_2-\overset{\overset{\displaystyle O}{\|}}{C}-OH$	Aspartic acid
	$-CH_2-CH_2-\overset{\overset{\displaystyle O}{\|}}{C}-OH$	Glutamic acid
Basic R-Group	—CH$_2$—CH$_2$—CH$_2$—CH$_2$—NH$_2$	Lysine
	$-CH_2-CH_2-CH_2-NH-\overset{\overset{\displaystyle NH}{\|}}{C}-NH_2$	Arginine

	—CH$_2$—C = CH
	  HN N
	   $\diagdown$C$\diagup$
	    $\mid$
	    H
		Histidine

solid lattice are greater than the energy benefit from dispersion in the melt or solution energy obtained upon dissolution. Cellulose is also found as approximately 50% of wood, together with lignin and other components. Wood of course has long been used as a structural material.

These natural polymers are flexible, but are not processable in solution or in the melt. A less common natural material which is genuinely processable and which became available to industrialized nations in the early 1800s was natural rubber.

DNA (Deoxyribonucleic acids)
Another class of natural polymer of great importance to life, but not useful for their mechanical properties are the nucleic acids which are used to transmit genetic information in living creatures, even bacteria. The basic polymer structure (Fig. 1.18) is a long chain of nucleotide bases attached to sugar rings joined by phosphodiesters.

A feature of DNA is its very high RMM. A human cell is some 20 μm in diameter of which the nucleus is part. The DNA, despite its carefully maintained helical nature over a short distance, is able to coil and fold over long distances, as most polymers do. If the DNA from a single human cell nucleus is uncoiled, it is over 1 metre long (in chromosomal fragments of up to 20-cm individual lengths), an astonishing example of polymer dimensions. Its cross-sectional area of course is very narrow

Natural rubber
This is obtained as latex (a heterogeneous aqueous suspension of micro-globules) from the sap of a tree. (The name originates from early use as an eraser.) After solving some initial processing difficulties (the material needed to be ground up to improve flow to allow ease of fabrication), Mackintosh made use of its elastic properties and developed rainproof clothing in the early 1820s.

However, a drawback to greater commercial acceptance was the temperature performance of the rubber and also the stickiness of the polymer in warm conditions which necessitated not only a textile lining for a coat but also a textile outer surface. This doubled a labour-intensive part of the process and mitigated against cost-effectiveness of garments made from this new material.

In 1836 in America, Charles Goodyear addressed the problem of surface tackiness and tried heating with elemental sulphur. What he obtained was a considerable increase in strength and mechanical properties. The process, later termed 'vulcanization', produced dramatic effects and formed the basis of the vehicle tyre industry, giving an impetus to the eventual development of motor transport among other things. Strength is increased by joining polymer chains together to give a giant molecule, linked by polysulphur bridges. The chemistry is somewhat complex and is elaborated in Chapter 5.

Cellulose esters
Cellulose is a triol (Fig. 1.18) containing one primary and two secondary alcoholic hydroxyl groups. By treating cellulose as an alcohol, workers in the mid-to-late-1800s identified routes of producing processable cellulose, some of which are still used on an appreciable scale today.

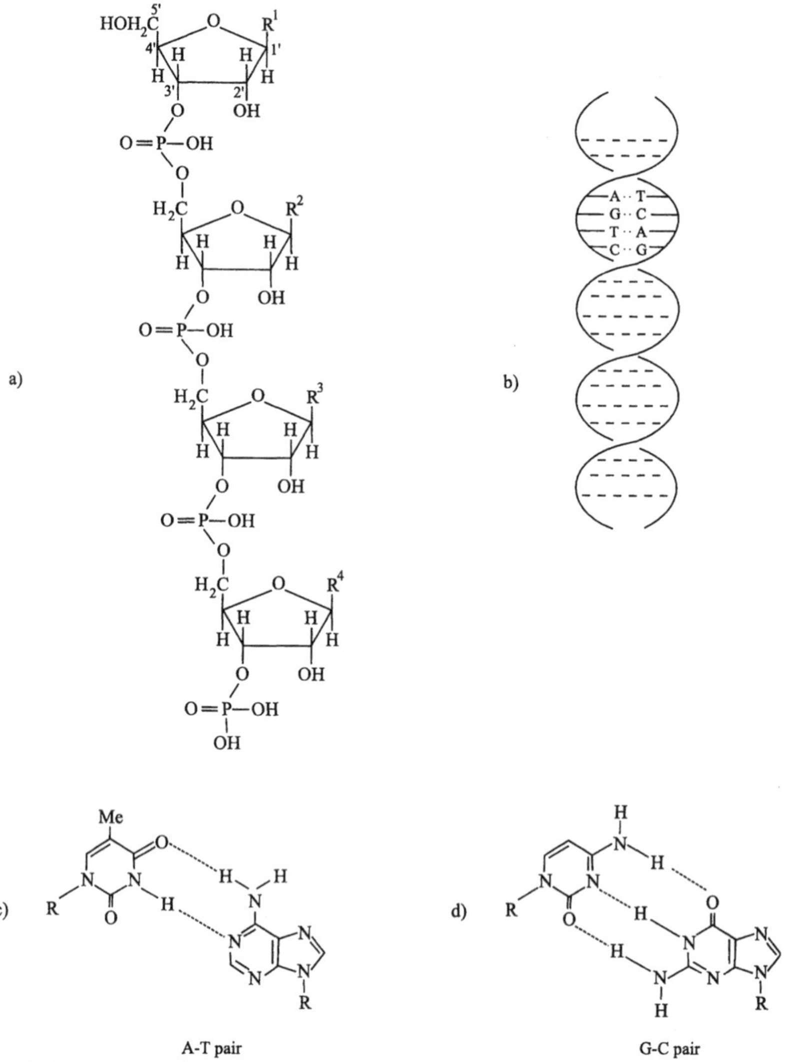

Fig. 1.18 Natural polymer nucleic acids:
a) a tetranucleotide-4 units of ribonucleic acid (RNA)
b) double helix structure of deoxyribonucleic acid (DNA)
c) base pairing, adenine-thymine
d) base pairing, guanine-cytosine

The basic strategy was to diminish hydrogen bonding by selective conversion of alcohol groups into esters of different types. While esters of organic alcohols with mineral acids (e.g. HNO_3) are not commonly used in organic chemistry, being generally more reactive than esters of an alcohol with a carboxylic acid, Schonbein showed in 1848 that cellulose treated with nitric acid in the presence of sulphuric acid was converted to nitrate esters (Fig. 1.19). These are sometimes called "nitrocellulose". This is a misnomer because they contain C-O-N bonds rather than the C-N bonds found in true nitrocompounds.

The most important feature of the nitration reaction is that it can be controlled. Complete reaction produces cellulose trinitrate, a solid known as

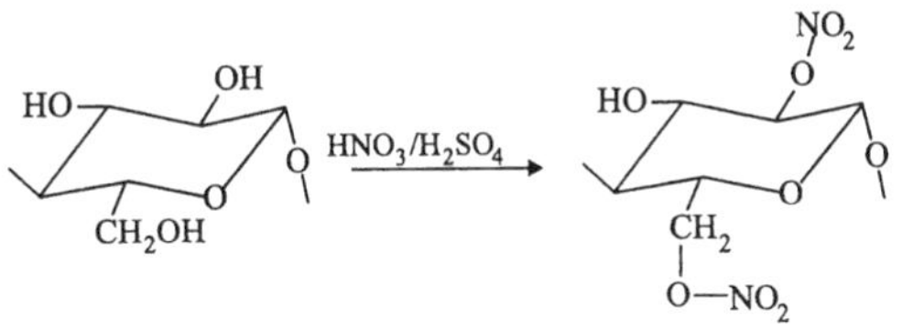

Fig. 1.19 Representative formation of cellulose nitrate

'gun cotton' and which is a shock-sensitive explosive. A lesser degree of nitration is achieved by increasing the water content of the reaction mixture. It is also possible by deliberate further heating to cut the long natural chains and give a lower RMM material which is soluble in ether/ethanol. This product was known as 'collodion', a lacquer which could be used medicinally by dabbing on to wounds and allowing the solvent to evaporate to leave an 'artificial skin'. It was also used in the early development of photography to coat the glass plates that were then used. Cellulose nitrate was arguably the material at the birth of the polymer industry, when Parkes demonstrated solid moulded articles from mixtures of it with oils at the Great International Exhibition in London in 1862.

A further use of cellulose nitrate was developed in 1884 by Le Chardonnet, who opened up the field of 'regenerated' cellulose, a strategy in which the natural polymer is converted to a processable intermediate, here the nitrate ester, and is then turned back to cellulose again by reduction with ammonium hydrogen sulphide ($NH_4^+HS^-$). The fibres so obtained have subtle differences from the original cellulose, in particular a silken lustre. The 'Chardonnet silk' immediately found a high added-value application in the fashion industry and continued to provide an artificial alternative to expensive natural silk until the 1930s when clothing from fully artificial polymers ('nylon' polyamides and later 'terylene' polyesters) became widely available throughout society.

Arguably the most important cellulose nitrate derivative, at least in terms of its effect upon the development of modern life, was the discovery that a mixture of solid cellulose dinitrate and camphor was melt-processable. (Melt-processing is a particularly useful technology which here allows production of optically transparent films as continuous strips.) The material was called celluloid and it revolutionized the photographic industry which until then was forced to rely on rigid, inflexible and breakable glass plates to support the necessary light-sensitive chemicals. The availability of this tough, flexible, pliable, and transparent material which could be transported around sprockets and spools allowed the development of the film industry with its great effect on twentieth century life. However, being composed of a nitrate ester, the early movie films were potentially flammable and presented a fire hazard to early cinemas. Although celluloid is no longer used, it has an important place in polymer history as being a truly melt-processable derivative of otherwise intractable cellulose.

Viscose Rayon
The success of Le Chardonnet silk prompted other approaches to regenerated cellulose. For example, the viscose Rayon process of Cross and Bevan in 1892

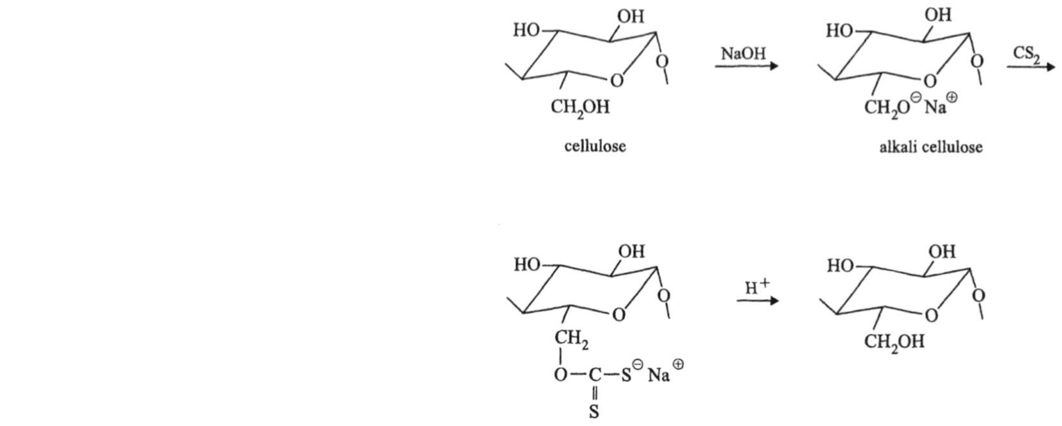

Fig. 1.20 The viscose rayon process

employed as the first step the treatment of cellulose (here behaving as a typical alcohol) with a base to form a water-soluble alkoxide salt, a step which is also now used to make a number of permanently modified cellulose derivatives (Fig. 1.20). Cross and Bevan reacted the alkoxide (alkali cellulose) with carbon disulphide to give a xanthate salt (Fig. 1.20). The resulting viscous solution was then made acidic and forced through spinnerets (or another processing device) where it lost carbon disulphide to give regenerated cellulose either as fibres, Rayon, or as transparent sheets of Cellophane with essentially the original structure. This was a substantial processing enhancement since the material could be recovered with slight effect on the original RMM. The processing advantage for Rayon was that the fibre thickness and dimensions could be controlled, whereas cotton fibres possess dimensions dictated by the plant on which they grow.

Since carbon disulphide is both toxic and extremely inflammable, its choice in the above process may not seem obvious. Bicarbonate might seem more obvious since at first sight the chemistry is similar (Fig. 1.21). However, bicarbonate chemistry is too unstable to be useful in this application. The eventual choice of system shows how commercial chemical products require substantial developmental studies after the original idea has been shown to be viable.

Also, in the case of cellophane the actual process contains a number of stages, all optimized after considerable effort. This is a feature of all industrial processes, but polymer processes show this particularly well, and will be emphasized in this primer. Thus, although the underlying chemistry may seem simple in concept, the actual practical system may be more sophisticated than otherwise expected.

a) $\quad HO^{\ominus} + CO_2 \longrightarrow HCO_3^{\ominus} \xrightarrow[H^{\oplus}]{heat} H_2O + CO_2$

b) $\quad RO^{\ominus} + CS_2 \longrightarrow ROCS_2^{\ominus} \xrightarrow[H^{\oplus}]{heat} ROH + CS_2$

Fig. 1.21 Comparison of a) bicarbonate & b) xanthate chemistry

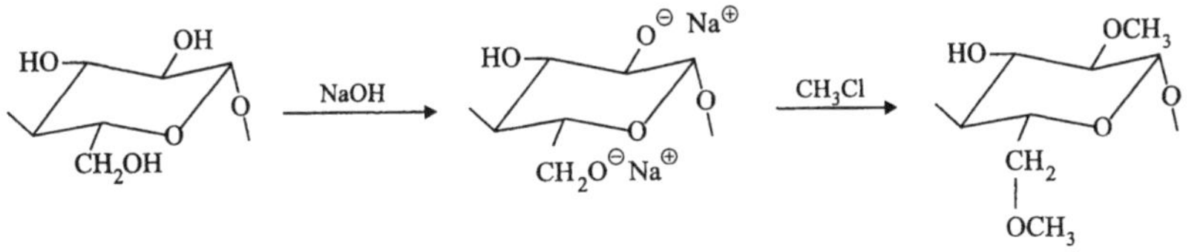

Fig. 1.22 Synthesis of Methyl Cellulose (a water-soluble derivative) (if ClCH₂CH₂OH replaces CH₃Cl then hydroxyethyl cellulose is obtained, in which -OCH₂CH₂OH replaces -OCH₃ in the product)

A whole range of cellulose polymers is now available using the initial step of the xanthate process. For example, if 2-chloroethanol replaces carbon disulphide then hydroxyethyl cellulose, a hydrophilic polymer used as a filler in ice cream is produced, while the use of chloromethane leads to methyl cellulose (Fig. 1.22). These products are sometimes termed 'cellulose ethers', and, although cellulose itself is a polyether, the name refers to the addition of the extra ether functionality.

Cellulose acetate
Schutzenberger first attempted the esterification of cellulose in 1865 but found the reaction too difficult to control. Treatment with acetic acid and acetic anhydride led to the formation of the triacetate (Fig. 1.23), a product which has relatively poor mechanical properties and, more importantly at the time, is only soluble in halogenated solvents such as methylene chloride (dichloro-methane). At the end of the last century these solvents were very expensive and could not be countenanced for large-scale applications. Nowadays these solvents are cheaper and the economics are overturned such that cellulose triacetate is used for cigarette filters.

However, a more useful material is the diacetate, in which (approximately) two hydroxyl groups are acetylated per ring. This leaves one hydroxyl group for hydrogen bonding. It was discovered in the early 1900s that the diacetate could be made in a two-step process in which acetylation with acetic anhydride containing sulphuric acid catalyst gives the triacetate, then controlled partial hydrolysis with dilute aqueous sulphuric acid gives the diacetate. This exploits the discovery that while acetylation cannot be controlled, hydrolysis of

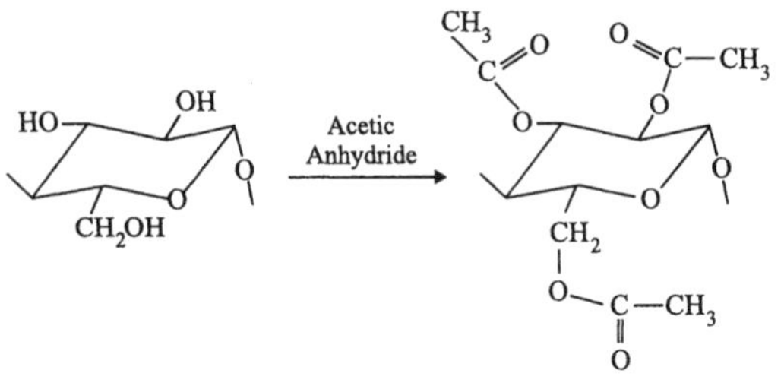

Fig. 1.23 Formation of cellulose triacetate

the acetate esters can. Cellulose diacetate is soluble in acetone, which was then, and still is, one of the cheapest solvents. One of the reasons that cellulose diacetate rapidly became accepted was that it could be used to strengthen canvas ('cellulose dope'). Thus an acetone solution was sprayed onto, for example, the canvas wings of the newly invented aeroplanes, and the solvent allowed to evaporate, giving an early example of the strengthening effect of a composite material. Military use in the First World War provided an impetus for acceptance of the new material.

Cellulose diacetate became the major thermoplastic moulding material in the early part of the twentieth century and is still used for some minor applications such as plastic combs and toothbrushes, although polyethene and polystyrene have now taken over in importance.

Cellulose materials continue to remain important. In the mid-1990s Courtaulds, who pioneered Rayon and cellulose acetate commercially, produced a new cellulose-based polymer ('Tencel') which employs a different solvent system. The benefit of cellulose is that it is inherently biodegradable and environmentally friendly; it is of course continuously renewable being obtained from plant sources. In principle it will still be available when coal- and petroleum-based resources become scarcer.

Artificial polymers

The remainder of this *Primer* mainly concerns artificial polymers — chain and step-growth. The distinction between chain polymerization and step-growth polymerization is crucial to the commercial preparations of these materials. Quite different conditions are employed and control of the reaction (to give the desired RMM and polymer properties) is therefore quite different.

The synthesis of artificial polymers (Tables 1.5 and 1.6) has actually been studied as long as the chemical modification of natural polymers. The photochemical polymerization of vinyl chloride and the polymerization of styrene were discovered respectively in 1838 and 1839 just as Mackintosh and Goodyear were improving the properties of natural rubber. However, this new polymeric state of matter was not really understood and the large scale commercial exploitation of polyvinyl chloride (PVC) waited almost 100 years until 1927 with polystyrene following in 1937. Of the step-growth polymers, nylon 6,6 became commercially available in 1938 with polyesters later.

Since then polymers have become more and more widely used. Exploitation of their physical and mechanical properties and developments in processing to open up further applications have all helped, but a recent development is in the use of functional polymers in which a property is exploited outside those due simply to the long chain structure. These include the presence of chemically reactive pendant groups, the conduction of electricity, biomedical uses of polymers and the like. These will be discussed towards the end of this *Primer*.

Although polymers tend to be thought of as modern materials, industry tends to use pre-IUPAC nomenclature which can be confusing. Accordingly, Table 1.5 gives old and new names, some acronyms and structures for many important chain-growth polymers which may be variously referred to throughout this book. Table 1.6 gives the structures of important step-growth polymers.

Table 1.5 Representative artificial chain-growth polymers

Industrial Names	Polymer Acronym	Monomer IUPAC Name	Repeat Unit
Polyethylene	(PE)	ethene	$+CH_2-CH_2+$
Polypropylene	(PP)	propene	$+CH_2-CH+$ with CH_3
Poly(vinyl chloride)	(PVC)	chloroethene	$+CH_2-CH+$ with Cl
Poly(vinyl alcohol)	(PVA)	cannot be made from monomer	$+CH_2-CH+$ with OH
Polystyrene	(PS)	phenylethene	$+CH_2-CH+$ with phenyl ring
Polyacrylonitrile	(PAN)	cyanoethene	$+CH_2-CH+$ with $C\equiv N$
Poly(vinyl acetate)	(PVA)	ethenylethanoate	$+CH_2-CH+$ with O, $C=O$, CH_3
Poly(acrylic acid)	(PA)	propenoic acid	$+CH_2-CH+$ with $C=O$, OH
Polyvinyl pyrrolidone	(PVP)	N vinyl pyrrolidone	$+CH_2-CH+$ with pyrrolidone ring (O, N)
Poly(methyl methacrylate)	(PMMA)	methyl (2 methyl propenoate)	$+CH_2-C+$ with CH_3, $C=O$, O, CH_3
Poly(vinylidene chloride)	(PVC)	1,1' dichloroethene	$+CH_2-C+$ with Cl, Cl
Polyacetylene	(PAC)	ethyne	$+C=C+$ with H, H

Table 1.5   *continued*

Industrial Names	Polymer Acronym	Monomer IUPAC Name	Repeat Unit
Polyisobutylene	(PIB)	2-methyl propene	$+CH_2-C+$ with CH_3 above and CH_3 below
Polyvinylidene fluoride	(PVDF)	1,1' difluoroethene	$+CH_2-C+$ with F above and F below
Polytetrafluoroethene	(PTFE)	1,1',2,2' tetrafluoroethene	$+C-C+$ with F F above and F F below
Polybutadiene (shown as trans isomer)	(BR)	buta 1,3 diene	$C=C$ structure
Polyisoprene (shown as cis form, natural rubber)	(NR)	2 methyl buta 1,3 diene	$C=C$ structure
Polychloroprene (shown as cis isomer)	(CR)	2 chloro buta 1,3 diene	$C=C$ structure
Polyvinylcarbazole	(PVK)	N vinylcarbazole	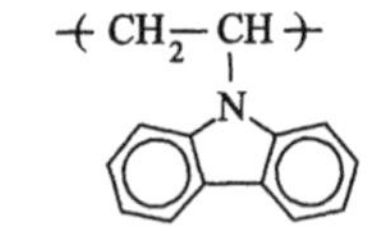 $+CH_2-CH+$ structure

Chain Copolymers

Acrylonitrile-styrene   (example)

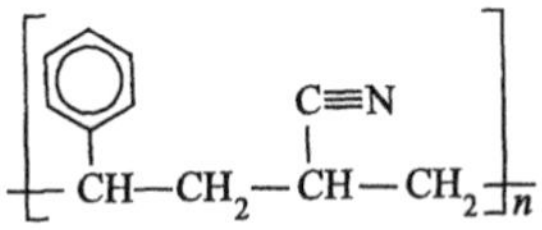

$$\left[CH-CH_2-CH-CH_2 \right]_n$$

(with benzene ring and $C \equiv N$ substituents)

numerous other systems include:-
acrylonitrile - butadiene - styrene (ABS)
styrene - butadiene (SBR)
ethylene - propylene (EPM/EPDM)
vinyl chloride - vinyl acetate etc.

Table 1.6 Representative artificial step-growth polymers

Class	Generalised Repeat Unit
Polyesters	
Polyamides	
Polycarbonates	
Polyurethanes	
Polysulphones	
Polyacetals	$\left(CH_2-O\right)$
Poly(phenylene oxide)	
Polyethylene oxide	$\left(OCH_2CH_2\right)$
Polyimides	
Polybenzoxazoles	
Polybenzimidazoles	
Polysiloxanes	
Polyetheretherketone	

Table 1.6   *continued*

Class	Generalised Repeat Unit
Polypyrrole (shown in uncharged form)	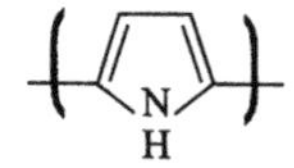
Polyaniline (shown in uncharged form)	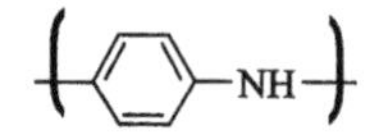

3 - Dimensional Network

| Phenol-formaldehyde (reacts at ortho & para positions) | |

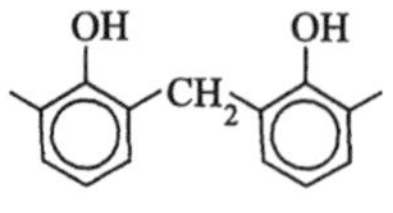

also includes ether links

Copolymer

| Polyether-polyamide numerous others:- for example polyether — polyurethane | |

1.6   General principles of industrial polymer synthesis

It is important to emphasize commercial and practical aspects since these underpin the great usefulness of polymers in modern society. Practical industrial preparations often differ from laboratory scale synthetic routes, a principle underlined in the polymer industry. In this *Primer* several practical polymerizations are examined in detail to give a feel for the factors which must be addressed in the development of feasible processes. However, first some general principles are discussed.

Economics

This can be very difficult to predict since each process has its own economics. For example, extreme situations could be the production of a large-volume low-cost product (e.g. polyethene) compared to a high added-value low-volume material (e.g. Kevlar^{TM}) where an expensive process can be tolerated.

Other costs include raw materials, e.g. solvents and reagents; oxygen in the air is by far the cheapest oxidant, water is much cheaper than any organic solvent; mineral acids and alcohols are relatively cheap, while ethanoic acid (and its derivatives) is the cheapest organic acid.

Ideal processes make use of all products, since the selling of an unwanted side-product can tip the economic balance of a process. Significant formation of a byproduct which is of no value represents a complete waste of mass which has to be moved, lifted, dissolved, pumped, filtered or otherwise taken through the production line: thus there is considerable impetus to improve any process. Often, however, this does not extend to implementing a completely new synthetic strategy because a major investment in new plant is then required.

Another important consideration nowadays concerns environmental aspects. The toxicity of reagents, byproducts, effluents, emissions, and general health and safety concerns can mitigate against an old process and enforce a new one. Environmental legislation has now become an important criterion in process selection, and can alter the traditional interplay of economic factors.

Catalysts

Although strictly a catalyst alters the rate of a chemical reaction without changing the position of equilibrium, the term tends to be applied to any species which improves the quality of a reaction. A yield improvement of a few per cent can alter the economics of a reaction and industry expends considerable effort to develop improved catalytic systems. Practical catalysts are often sophisticated mixtures with accelerators, promoters and other species present, with the exact detail being kept secret. In particular, the use of heterogeneous transition metal species is very common. The variable valency of the metal atom allows electron transfers to occur with concurrent changes of orbital geometry and other factors which are important in the chemistry of co-ordination complexes and organometallic systems. The catalyst is often solid in an otherwise liquid or gaseous medium, such that surface phenomena are controlling factors. Catalyst science is a large and increasingly important topic beyond the scope of this book, and virtually every optimized industrial chemical reaction involves a catalytic system. Polymer manufacture from initial raw materials through to final products shows the important of catalysts very well and there are a number of catalytically enhanced reactions which would not be considered on laboratory scale, e.g. the oxidative conversion of benzene to maleic anhydride in vapour phase over a vanadium pentoxide catalyst.

The cost of a specialized catalyst is offset by the relatively small quantity required and the number of times it can turnover before deactivation.

Purity

An important consideration in catalysis is 'poisoning' in which small traces of impurities deactivate the catalyst. Sulphur-containing species seem particularly deleterious to a wide range of catalytic systems. It is always desirable in any case to avoid impurities in a reaction and in polymer manufacture particularly so, because of the effect of trace impurities upon the final polymer molar mass. A species which deactivates the propagating end-group in a chain polymerization can significantly alter RMM distribution and a variable batch-to-batch impurity level can complicate calculation of the appropriate initiator concentration. As will be seen later in step-growth polymerizations, the final RMM relates to the degree of polymerization such that for the joining of 10 000 units in a chain, an extent of reaction of 99.99% is required (see Chapter 2). Imagine being asked routinely to produce a yield of 99.99% in a typical laboratory experiment! Clearly high purity is crucial here, as well as are other means to drive the reaction towards maximum yield. Industrial strategies to do so sometimes involve reaction conditions that would not be employed in the small-scale laboratory.

General reaction conditions

The small-scale chemical laboratory typically employs reflux in glass-walled apparatus as the major methodology for a thermally-activated chemical process. This is thermodynamically a constant pressure (1 atmosphere) and variable volume system in which the equilibrium constant depends upon Gibbs Free energy; but it should be recalled that there is another thermodynamic regime at constant volume and variable pressure. This is harder to implement in the laboratory but industry is at home with constant volume systems, using stainless steel reactors or other infrastructure which can tolerate extremes of high or low pressure. Thus reactions which are inefficient in the laboratory can be driven over to effective thermodynamics by proper choice of conditions. A good example of a reaction that would be difficult and hazardous in the laboratory, but which is industrially effective, is hydrogenation with hydrogen gas at several hundred atmospheres' pressure.

Process enhancements

These are sought at all levels and continual effort goes into better economics and process improvement. Thus even in a well established polymer system a new synthetic reaction, improved catalysts and diminution of side-products and waste are all axiomatic benefits. Both heating and cooling cost money and any development which minimizes extremes of temperature is very much sought after. Better control of polymer RMM (thus physical properties) is a key issue, while a whole science outside the scope of this book concerns engineering aspects of polymer processing and how these affect polymer applications and performance.

2 Polymer properties and characterization

2.1 Molar mass

Typically vinyl polymers (from chain polymerization of alkenes) have molar masses (RMMs) in the range of 10^5–10^6 whereas typical step-growth condensation polymers (e.g. polyamides or polyesters) may be as low as 15 000–20 000. The techniques most commonly used to determine polymer RMM include end-group analysis, osmometry, light scattering, ultracentrifugation, sedimentation, viscometry and chromatography. However, most of these involve rather lengthy procedures and in practice molar masses are obtained from high performance gel permeation chromatography (HPGPC) or viscosity measurements. It is important to recognize that the fundamental measurements of molar mass must be performed on dilute solutions so that intermolecular interactions can be ignored.

Number average molar mass

End-group analysis

The number average molar mass ($\overline{M_n}$) can be measured for any polymer which has an end-group that can be determined by physical or chemical means. However, it must be remembered that the concentration of end-groups in a polymer is low and this will set a practical (experimental) limit to the measurement. Take for example an unsaturated polyester as used in the manufacture of fibre and film reinforcement which is usually of fairly low molecular weight and is hardened by subsequent chemical reactions (i.e. cross-linked) (see 'fibreglass' Chapter 5). To determine M_n one can titrate the terminal COOH and then determine the moles of polymer per gram using eqn 2.1 and then apply eqn 2.2.

$$\text{Moles polymer/g} = \frac{\text{Moles COOH (or moles OH)}}{\text{Sample mass}} \tag{2.1}$$

$$\overline{M_n} = \frac{1}{\text{Moles polymer per gram}} \tag{2.2}$$

Membrane osmometry

In simple terms osmosis can be compared to diffusion except that in diffusion movement takes place from high to low pressure (or concentration), whereas in osmosis movement takes place from a region of high chemical potential (pure solvent) to one of lower chemical potential (solution) through a membrane. In practical terms the experiment is easier to perform by applying a pressure to the solution compartment of the cell to stop the osmotic action. Osmotic pressure (Π) is related to the polymer solution concentration (c) and molar mass ($\overline{M_n}$) by the Van't Hoff equation (eqn 2.3)

$$\left(\frac{\Pi}{c}\right)_{c=0} = \frac{RT}{\overline{M_n}} + A_2 c \tag{2.3}$$

where A_2 is a constant and c is the concentration the polymer solution. A plot of $\dfrac{\Pi}{c}$ *versus* c, when extrapolated to $c = 0$, has an intercept of $\dfrac{RT}{\overline{M_n}}$

The major source of error is due to low molar mass species diffusing through the membrane and the technique is most useful over the RMM range 50 000–2 000 000.

Vapour pressure osmometry
This technique is most useful for molecular masses below 25 000. The principle is analogous to membrane osmometry although there is no membrane. A drop of solvent and a drop of solution are placed by syringe on matched thermistors in an insulated chamber saturated with solvent vapour. Condensation heats the solution thermistor until the vapour pressure is increased to that of pure solvent (eqn 2.4).

$$T = T_s - T_o = -\frac{RT_o^2}{\Delta H_v} m \tag{2.4}$$

where T_s and T_o are the temperatures of the solution and solvent respectively, m is the molarity of the solution (moles/1000 g of solvent) and ΔH_v is the molar enthalpy of vaporization.

Experimentally the temperature change (ΔT) is measured by the change in resistance between the two thermistors ($\Delta R = k_R \Delta T$) for a series of solution concentrations (c, g/dm^3). A graph of $\Delta R/c$ *versus* c (eqn 2.5) is constructed, the intercept of which allows $\overline{M_n}$ to be deduced.

$$\Delta R/c = K\left[\frac{1}{\overline{M_n}} + Ac\right] \tag{2.5}$$

Mass average molar mass

Light scattering
This technique depends on a measurement of turbidity which is the proportion of light removed from the primary beam by scattering as it passes through a layer of solution.

Experimentally, the ratio of intensity of scattered light (i_θ) to the intensity of incident light (I_θ) is measured at a particular distance (r) from, and at an angle θ to the source. The result is converted into the Rayleigh ratio, R_θ ($r^2 i_\theta/I_\theta$) and thence to turbidity Γ (eqn 2.6). Finally, the molar mass is obtained using the Debye Equation (eqn 2.7), where a plot of $\dfrac{Hc}{\Gamma}$ *versus* c gives an intercept of $\dfrac{1}{\overline{M_w}}$.

$$\Gamma = \left(\frac{16\pi}{3}\right)\left(\frac{R_\theta}{1 + \cos\theta}\right) \tag{2.6}$$

$$\frac{H_c}{\Gamma} = \frac{1}{M_w} + Bc \qquad (2.7)$$

Here B is a constant, c is the concentration of the polymer solution, H is a function (eqn 2.8) of the wavelength (λ) of light, the refractive index of the solvent (n) and the change in refractive index with concentration ($\frac{dn}{dc}$).

$$H = \left(\frac{32\pi^3}{3N\lambda^4}\right)n^2\left(\frac{dn}{dc}\right)^2 \qquad (2.8)$$

Sedimentation
This method is based upon the fact that a particle of mass m $(\overline{M_w}/N)$ and specific volume V_p, when placed in a solvent of density ρ_s will:

1   fall due to the influence of gravity (mg);
2   experience an upthrust due to Archimedes principle, the upthrust being equal to $V_p\rho_s$;
3   be subject to a viscous drag as it falls i.e. FU, where F is the frictional coefficient (equal to $6\pi\eta r$, Stokes Law) and U is the particle velocity (r is the radius of the particle and η is the viscosity of the medium).

The particle will fall slowly at first, accelerating until it reaches a constant velocity called the **terminal** or **sedimentation** velocity, U_s. At this point the 'downward' forces equal the 'upward' forces (eqn 2.9), from which $\overline{M_w}$ (equal to mN) can be determined.

$$mg = V_p\rho_s g + FU_s \qquad (2.9)$$

Using a centrifuge allows larger values of g to be employed.

Ultracentrifugation
This is the most intricate and expensive method for determination of M_w. It is not used much for synthetic polymers but mainly for natural polymers (e.g. proteins). It is based upon the principle that molecules, under the influence of a strong centrifugal field (several thousand times the acceleration due to gravity), distribute themselves according to size perpendicularly to the axis of rotation. The progress of the solute under the influence of the field can be followed optically and allows measurement of the sedimentation velocity, which in turn allows the deduction (eqn 2.10) of the sedimentation constant (s), a parameter necessary for the determination of the molar mass (eqn 2.12).

$$s = \frac{m(1 - V_p\rho)}{f} \qquad (2.10)$$

$$D = \frac{kT}{f} \qquad (2.11)$$

$$\frac{D}{s} = \left(\frac{RT}{M_w}\right) \times \frac{[1 + Ac]}{(1 - V_p\rho)} \qquad (2.12)$$

Here m is the mass of the particle of partial specific volume V_p immersed in a solvent of density ρ, and D and f are the diffusion constant and frictional force, respectively.

High performance gel permeation chromatography

This is the most convenient method of obtaining both $(\overline{M_n})$ and $(\overline{M_w})$. The polymer solution is injected into a solvent stream which flows through a column packed with a highly porous material which separates the polymer molecules according to their size (size exclusion chromatography, SEC). The small molecules enter the pores and are retarded relative to the larger one. Detection of the polymer in the eluant is usually by either refractive index (RI) or ultraviolet/visible spectroscopic detectors. The instrument needs to be calibrated with samples of known RMM. This can sometimes be problematic if a polymer of a new type is being studied because available standard RMM samples may be from a different polymer type. Polystyrene is most often used as standard. Ideally, the standard samples should be from the same polymer as the test sample.

Viscometry

Absolute methods (e.g. light scattering) for $\overline{M_w}$ are somewhat difficult and time-consuming and viscometry provides a faster and cheaper method. Viscometry operates on the principle that on dissolution in a solvent, a polymer molecule increases its dimensions and becomes more viscous and hence the slower its movement or flow. Take, for example, the addition of methyl cellulose (wall paper paste) to water — there is a considerable increase in viscosity.

Viscosity measurements are usually performed using fairly simple and basic equipment, e.g. a glass (Ubbelhode) viscometer (Fig. 2.1); the extra tube in the apparatus is for pressure equalization as the liquid level changes. Experimentally, one would measure the flow times (t_s) for several concentrations (c) of polymer solution (maximum concentration approximately 0.5 g/100 cm^3), determine the specific viscosity η_{sp} (equal to $(t_s - t_o)/t_o$, where t_o is flow time of pure solvent) for each solution concentration and plot (η_{sp}/c) *versus* c to yield a straight line whose intercept will allow the determination of the molar mass using eqns 2.13 and 2.14.

$$\frac{\eta_{sp}}{c} = [\eta] + A[\eta]^2 c \tag{2.13}$$

A is a constant (0.35–0.40) for a series of polymer homologues in a given solvent, and $[\eta]$ is known as the 'intrinsic' viscosity (eqn 2.14).

$$[\eta] = KM^a \tag{2.14}$$

Here K and a are constants for a particular polymer–solvent system and can be found by using known standards (from light scattering) and fitting the data to eqn 2.15.

$$\log[\eta] = \log K + a \log M \tag{2.15}$$

Viscosity average RMMs lie between those corresponding to M_w and M_n but are usually closer to the M_w because of the greater effect on viscosity of larger molecules. In general the behaviour of polymer solutions shows a number of thermodynamic effects because of the large size of the molecules, and the full description of the behaviour of polymer solutions (which can be non-Newtonian) is beyond the scope of this *Primer*.

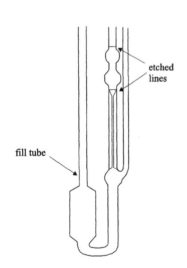

etched lines

fill tube

Fig. 2.1 Schematic of an Ubbelhode viscometer

It is worth mentioning a new class of polymers, 'star polymers' or dendrimers, in which a polyfunctional material grows from a central point branching in all directions to give an almost spherical high-RMM molecule matter like a dandelion 'clock' seedcase in shape (Fig. 2.2). These have interesting physical properties, including their behaviour in solution.

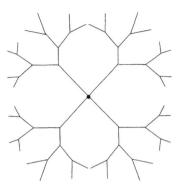

Fig. 2.2 Schematic of a 'star polymer' or Dendrimer

2.2 Polymer stereochemistry

As well having different lengths, many macromolecules have more than one structure for a given molar mass. This is easily visualized if one considers the homologous series of the alkanes. As Fig. 2.3 shows there is a simple progression of the linear chain from methane to propane.

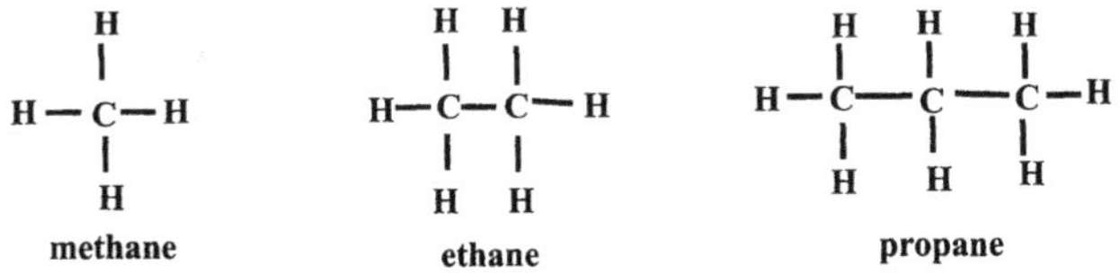

$$
\begin{array}{ccc}
\overset{\displaystyle H}{\underset{\displaystyle H}{H-C-H}} &
\overset{\displaystyle H\quad H}{\underset{\displaystyle H\quad H}{H-C-C-H}} &
\overset{\displaystyle H\quad H\quad H}{\underset{\displaystyle H\quad H\quad H}{H-C-C-C-H}} \\
\text{methane} & \text{ethane} & \text{propane}
\end{array}
$$

Fig. 2.3 Series of alkanes.

However, at butane we get the appearance of a phenomenon called **isomerism** (Fig. 2.4), i.e. a linear and a branched structure which both fit the molecular formula C_4H_{10}.

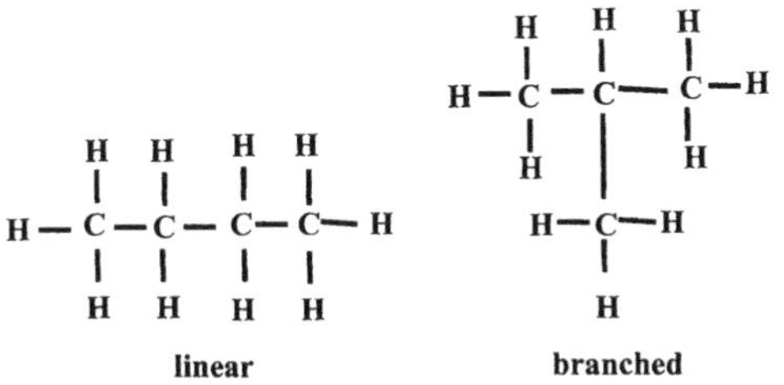

linear   branched

Fig. 2.4 Isomerism of butane.

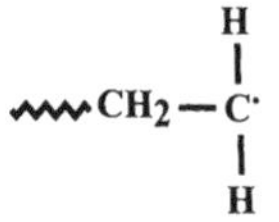

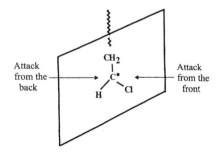

$$\text{www} CH_2 - \overset{\displaystyle H}{\underset{\displaystyle H}{C^{\cdot}}}$$

Fig. 2.5 Polymer radical.

Obviously, in the polymer situation, i.e. addition of say another 1000 CH_2 groups, an enormous number of **structural isomers** can be imagined. Even if we could produce molecules of the same RMM, they certainly would not have the same structure.

Not only are there structural isomers, but there are also stereoisomers and geometric isomers. Stereoisomers (isomers with different spatial arrangements) are produced as a consequence of monomer attack on the sp^3 active group involved in the polymerization reaction. Perhaps the best way to demonstrate the various possible different structural arrangements is to consider the addition of a monomer to the radical polymer chain shown in Fig. 2.5.

In Fig. 2.6, the monomer is a substituted (vinyl) alkene, $CH_2 = CHX$, exemplified for X = Cl, and the groups wwwCH_2, **H** and **X** of the polymer radical are arranged trigonally (120°) around the carbon radical ($C^{\bullet}$) such that attack from the incoming monomer can either be from the front or the back,

Attack from the back   Attack from the front

Fig. 2.6 Attack of monomer on polymer radical.

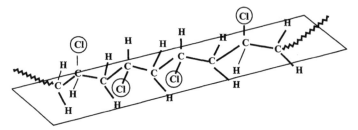

Fig. 2.7 Atactic spacial arrangement (groups arranged randomly).

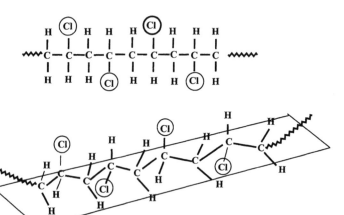

Fig. 2.8 Syndiotactic spacial arrangement (groups arranged alternately).

giving rise to three different spatial arrangements called atactic (Fig. 2.7), syndiotactic (Fig. 2.8), and isotactic (Fig. 2.9). These are best seen from Fischer projections of the chain.

For maximum crystallinity, polymer molecules, similar to small molecules (e.g. iodine), should be composed of arrangements or structures which can pack easily, i.e. isotactic or syndiotactic. However, some atactic polymers are also able give crystalline polymers, the best example being linear (non-branched) polythene which can pack in the solid lattice in a similar regular manner to small molecule material (Fig. 2.10).

It is also possible to get close packing in an atactic straight chain macromolecule even when some groups are larger than H. For example, in atactic polyvinylalcohol the hydroxy group is small enough for the polymer to pack into a lattice and a crystalline material results.

However, the majority of polymers are unlikely to produce straight chains naturally, unless the macromolecule is stretched, as it might be in an elastic band,

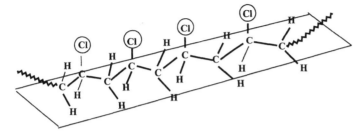

Fig. 2.9 Isotactic spacial arrangement (groups arranged on the same side).

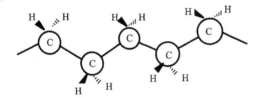

Fig. 2.10 Linear polyethene.

so that the likelihood of having 100% crystallinity is very low. Macromolecules are more likely to be coiled like cooked spaghetti and therefore the chances of close packing, and hence crystallization, are lessened. Any material with regular arrays can form crystalline regions, so isotactic and syndiotactic versions of chain-growth polymers are expected to be more crystalline in general. (In practice all vinyl polymer samples are either mixtures of isotactic and atactic molecules or mixtures of syndiotactic and atactic molecules. It is these differences in structure which dictate the extent of crystallinity or non-crystallinity (e.g. amorphous nature) found in a polymer.) The degree of ordering or crystallinity in a polymer can affect physical and mechanical properties, and control of structure is an important aspect of commercial polymerization procedures.

Most polymers tend not crystallize on cooling, instead they vitrify to produce a glass-like solid. To illustrate the significance of crystallization on vitrification, consider the volume–temperature curve for a low molecular weight compound such as glycerine (Fig. 2.11).

At high temperature (A), the glycerine is a somewhat viscous liquid. As the temperature is reduced the volume contracts in an approximately linear fashion until the melting point (T_m) is reached. At this point (B), isothermal crystallization occurs as more heat is removed. As crystallization proceeds, a sharp drop in volume accompanies it (B–E). When crystallization is complete, the temperature can fall again with thermal contraction of the crystalline solid (E–F).

On the other hand, no crystallization may occur. Glycerine is quite viscous at its melting point (T_m) and nucleation of the solid phase is slow. If no nuclei

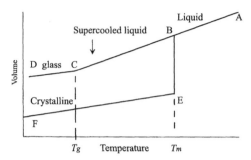

Fig. 2.11   Volume–temperature curve for glycerine.

are present, the glycerine may simply continue to cool as a (metastable) supercooled liquid (B–C) until such time as the viscosity is so high that flow is negligible (C–D). Below point C, the material is referred to as a 'glass' or a 'glassy solid'. Structurally, it is amorphous just as a liquid is but mechanically it acts as a solid. The temperature at which this occurs is called the **glass transition** temperature and is given the symbol T_g. T_g is as important a parameter in polymer technology as is the melting temperature, T_m.

For the final type of isomerism, geometric isomerism, to occur in a polymer the chain must contain double bonds. This is most commonly found where the initial monomer contains two double bonds (i.e. is a diene) so that on polymerization the chain still contains unreacted double bonds. Isoprene is an example (Fig. 2.12). The trans form packs better giving greater crystallinity and hence a harder rubber. It should also be noted that dienes can polymerise through one of the single double bonds (i.e. acting as a simple vinyl monomer) thereby giving rise to the various tacticities discussed above. For isoprene, the methyl group makes the carbons non-equivalent. Thus the 1,2 and 3,4 polymers may be represented in Figs 2.7, 2.8 and 2.9 by replacing Cl with $-CH=CH_2$ (for the 1,2 polymer) and $-C(CH_3)=CH_2$ (for the 3,4 polymer). For butadiene with no methyl groups, both vinyl polymer forms are the same (see page 81).

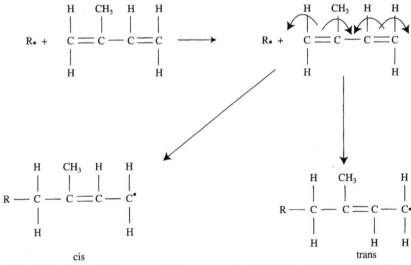

Fig. 2.12   Geometric isomerism in isoprene polymerisation (1,4 polymerisation shown only)

2.3 Structure–property relationships

The glass transition temperature

It has been shown previously that some liquids may be supercooled to form glasses (Fig. 2.11) without crystallization taking place as the temperature is lowered. In such materials, the viscosity (equal to the shear stress/velocity gradient) changes at so great a rate that over a small temperature interval the character of the material changes from a liquid to a rigid solid or glass. In fact for some polymers, in going from a liquid polymer melt or rubber to a rigid glass, the Young's Modulus (stress/strain) — a measure of the material's strength — may change by a factor of 1000. This change is now known to take place at the glass transition temperature, (T_g (see Section 2.6).

While the melting point of a polymer (T_m) represents a true phase change from solid to liquid and is governed by the laws of thermodynamics, the nature of the glass transition is not completely understood and is termed a second-order phase change since both phases before and after T_g are solids. However, since all polymers have glass transitions which affect the mechanical properties, it is convenient to treat the phenomenon as a true transition.

Glass transition temperatures can be determined by many experimental techniques, but traditionally, they are determined from the change in volume of the material as a function of temperature (e.g. Fig. 2.11).

Knowledge of the T_g value allows the generalization that a material with a T_g *above* room temperature is a plastic (i.e. it is rigid at room temperature) whereas a material with a Tg *below* room temperature is a rubber (i.e. it is flexible at room temperature).

Perhaps the best example of this phenomenon is the cooling of a rubber tube in liquid nitrogen. At room temperature the tube is flexible and elastic yet after several seconds of being immersed in liquid nitrogen it is stiff and brittle and can even be splintered into fragments with a hammer blow. This cannot be done if the tube is struck with the hammer at room temperature, where the energy of the blow is simply absorbed by polymer chain movement. The hammer rebounds from the rubbery solid. At a low temperature the chains cannot rotate or move and the impact breaks bonds in the polymer structure.

Another good example of the importance of the T_g can be underlined by considering polyvinyl chloride (PVC) (wet look, simulated leather) boots. These boots have a T_g of about +70 °C and are rigid and stiff at room temperature (25 °C) and are therefore impractical as footwear. However, by the addition of a plasticizer (which is a small molecule compound which improves flexibility; see Section 2.4.3) or the use of a copolymer of PVC (see Chapter 3), it is possible to lower the T_g to approximately +10 °C, thereby increasing the flexibility at room temperature. Unfortunately, the loss of plasticizer or the wearing of the boots at the lower temperatures found in the winter (1–2 °C) causes a return to rigidity with possibility of the material cracking.

In the absence of a plasticizer, T_g reflects inherent properties of the polymer chain in the solid. In the melt state any polymer molecule has both translational motion, in which the whole chain is moving, and chain segment movement, in which various sections (or segments) of the main chain of the polymer kink and unkink, with the rate of change of configuration being dependent upon the

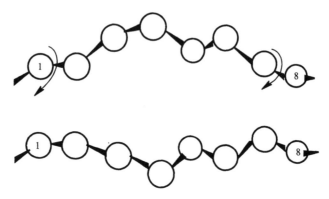

Fig. 2.13 Crankshaft motion.

temperature. To make sure that any macromolecule can undergo this internal movement there must be sufficient thermal energy to surmount the activation energy needed for the rotation around the chain bonds and also that there is a void or a hole in the total structure into which the particular segment of the chain can move. This is shown in Fig. 2.13, where the void into which the illustrated 8-carbon atom segment of the chain can rotate is known as the free volume and is temperature-dependent, such that the higher the temperature the bigger the void. The movement of the chain segment also depends on whether the polymer molecule has pendant groups attached to the main chain (or is branched), since such molecules will need larger free volumes if they are to rotate. Fig. 2.13 also shows that long molecules are not usually found as straight chains, but are twisted and kinked and hence there is less possibility of them fitting closely together to give crystals. In fact, it is this coiled nature that gives rubbers their elastomeric properties, because they can stretch out to give elongated straight chain polymers.

Since most polymers find application as solids it is important that the structural factors affecting the magnitude of the T_g are recognized and considered, since by doing so we can design molecules with appropriate temperature-dependence of properties.

Relation of glass transition to structure: factors affecting glass transition temperature

Chain flexibility
Chain flexibility is probably the most important factor in determining the magnitude of T_g. A consideration of the T_g values for polythene and polycarbonate (Fig. 2.14) can be rationalized in terms of increased rigidity of the chain, giving less opportunity for bond rotation and therefore an increase in T_g.

However, it is often difficult to separate the effect upon T_g of chain flexibility, compared to for example steric hindrance from the bulkiness of a side group.

Steric hindrance
This effect can be best illustrated by comparing the methyl-substituted styrenes. Having the methyl group in the ortho position gives an increase in T_g

poly(ethene)

$T_g = -120^0C$

polycarbonate

$T_g - + 150^0C$

Fig. 2.14 Effect of flexibility on glass transition temperature.

of 24 °C. This is because rotation of the back bone has been restricted, and T_g is increased as less 'cooling' is required to 'freeze out' motion. A similar comparison can be made for polymethylacrylate and polymethylmethacrylate (Fig. 2.15).

Side group effects
Any increase in the size of the group usually also causes steric hindrance and results in an increase in T_g. This is illustrated by comparing the series polythene, polypropene, polystyrene and polyvinylcarbazole in which the side groups are H, methyl, benzyl and carbazole (Fig. 2.16).

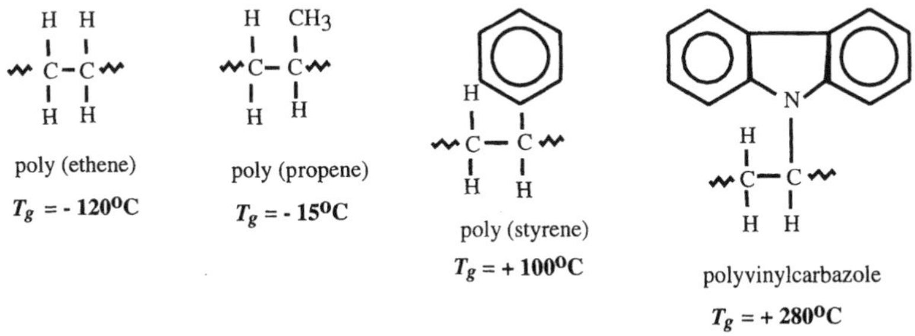

poly (ethene)
$T_g = -120^0C$

poly (propene)
$T_g = -15^0C$

poly (styrene)
$T_g = +100^0C$

polyvinylcarbazole
$T_g = +280^0C$

Fig. 2.16 Rigid side groups effect on T_g.

However, T_g depends not only on the size of the side group but also on its flexibility. For instance, in the series of acrylates (Fig. 2.17) the T_g actually decreases as the side group gets larger.

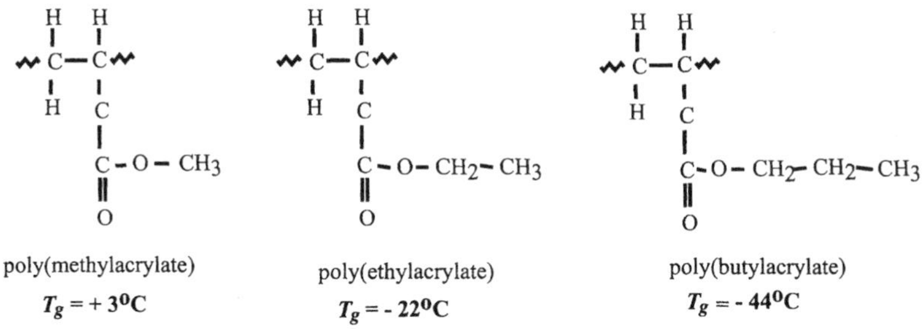

poly(methylacrylate)
$T_g = +3^0C$

poly(ethylacrylate)
$T_g = -22^0C$

poly(butylacrylate)
$T_g = -44^0C$

Fig. 2.17 Flexible side groups.

The increased flexibility of the side group compensates for its increased size and gives a more flexible polymer (i.e. a lower T_g owing to an increase in the free volume). Such action is often referred to as *internal plasticization*.

Symmetry
Any increase in symmetry lowers T_g. Typical examples are asymmetric PVC and the symmetric analogue polyvinylidene chloride (PVC_2) and, similarly, polypropene and polyisobutene (Fig. 2.18). In both cases the increased number of side groups is more than compensated by the increased symmetry.

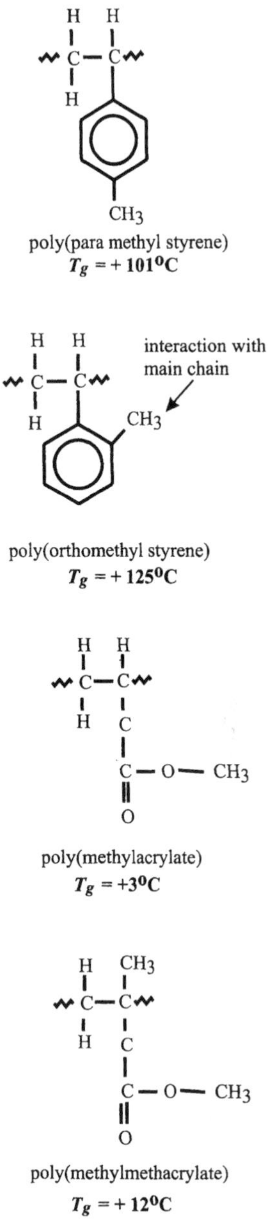

poly(para methyl styrene)
$T_g = +101^0C$

poly(orthomethyl styrene)
$T_g = +125^0C$

interaction with main chain

poly(methylacrylate)
$T_g = +3^0C$

poly(methylmethacrylate)
$T_g = +12^0C$

Fig. 2.15 Effect of steric hindrance on glass transition temperature.

polyvinylchloride
$T_g = + 87^0C$

polyvinylidene chloride
$T_g = - 17^0C$

polypropene
$T_g = - 14^0C$

polyisobutene
$T_g = - 65^0C$

Fig. 2.18 Symmetry and asymmetry effects on T_g.

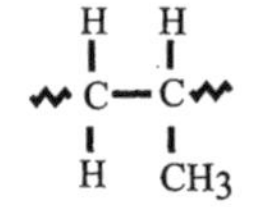

polypropene
$T_g = - 14^0C$

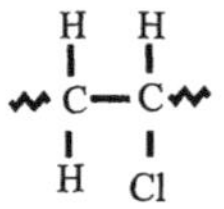

poylvinylchloride
$T_g = + 87^0C$

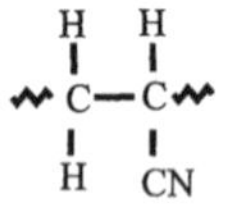

polyacrylonitrile
$T_g = + 103^0C$

Fig. 2.19 Polarity effects on T_g.

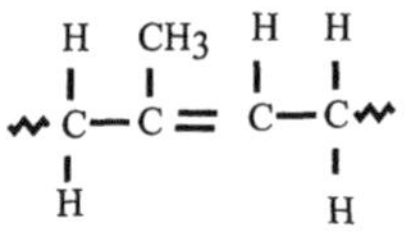

polyisoprene

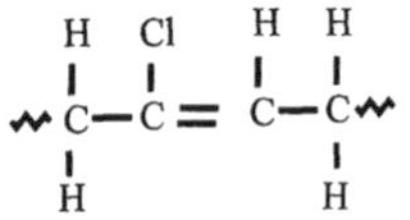

polychloroprene

Fig. 2.20 Structures of polymers from isoprene and chloroprene.

The most plausible explanation is a free-volume effect. Here the chains find it more difficult to pack, which leads to the presence of a larger free volume and therefore less thermal energy (i.e. lower temperature) is required to create sufficient free volume to allow chain rotation and hence flexibility.

Polarity

Any increase in polarity (or cohesive energy density) has a tendency to raise the T_g. For example, non-polar polypropene has a T_g below room temperature, moderately polar PVC has a T_g in excess of room temperature and highly polar polyacrylonitrile has a T_g in excess of 100 °C (Fig. 2.19). The increase in T_g can be rationalized in terms of the increased amount of polar bonding which occurs between the chains as the methyl group is successively replaced by the chlorine and cyano groups, respectively, thereby giving stronger bonding and restricting rotation about the backbone.

In a similar manner, replacing the methyl (CH_3) group in polyisoprene by the polar chlorine atom group (Cl) in chloroprene raises the T_g 25 °C from –73 °C to –50 °C (Fig. 2.20). Again this increase is due to restriction of rotation and movement of individual atoms by polar bonding.

Copolymerization

Although the concept of copolymers will be discussed further in Chapters 3 and 4 it is appropriate to mention here that copolymers have T_g's which are intermediate between those of the pure homopolymers. For example, for a copolymer consisting of polymer 1 (T_{g1}) and polymer (T_{g2}), then the resultant T_g (T_{gr}) is given by either eqn 2.16 or 2.17.

$$T_{gr} = V_1 T_{g1} + V_2 T_{g2} \qquad (2.16)$$

$$\frac{1}{T_{gr}} = \frac{W_1}{T_{g1}} + \frac{W_2}{T_{g2}} \qquad (2.17)$$

where the T_g's of homopolymers are given in Kelvin, V_1 and V_2 are volume fractions in the copolymer, and W_1 and W_2 are weight fractions of polymers 1 and 2 in the copolymer.

The use of these equations is best illustrated for the copolymer produced between styrene ($T_g \sim +100$ °C) and butadiene ($T_g \sim -70$ °C). This is the artificial synthetic rubber BUNA-S (also called BR rubber) which combines the hardness and mechanical stability of styrene and the flexibility of butadiene. By varying the amounts of each monomer copolymers can be

Table 2.1 Effect of monomer composition on copolymer T_g

Styrene (%)	Butadiene (%)	T_g	Nature of material at 55 °F
52	48	15 °C (59 °F)	Rigid elastomer
50	50	12 °C (54 °F)	
48	52	9 °C (48 °F)	Flexible plastic

produce with differing T_g's. Thus by careful manufacture we can produce goods which appear different (Table 2.1)

Crystallinity and melting point

Examples of crystalline polymers are high-density polyethene (HDPE), isotactic polypropene, 6,6-nylon, and polyvinylidene chloride. Semi-crystalline polymers behave like mixtures of amorphous (T_g) and crystalline (T_m) material, i.e. the polymer melts over a temperature range rather than possesses a sharp melting point (Fig. 2.21).

There are two useful empirical rules of thumb regarding crystalline polymers. The first relates the T_g to the melting point. For an unsymmetrical polymer such as PVC, the ratio $T_g/T_m \cong 0.66$, while for a symmetrical polymer such as polyvinylidene chloride ((poly(1,1 dichloroethene), PVC_2) the ratio $T_g/T_m \cong 0.5$, where T_g and T_m are in Kelvin. This identity seems to apply irrespective of whether polymers are chain-growth or step-growth. Thus for polyethylene terephthalate, which is unsymmetrical (Fig. 2.23) and has a T_m of 267 °C (540 K), the predicted T_g of 87 °C compares favourably with the measured value of 80 °C. For polyvinylidene chloride the agreement is not quite as good, with a T_g of −19 °C (254 K), giving a predicted T_m of 235 °C as compared to 190 °C. However, the pure homopolymer is relatively unstable which complicates the assessment of thermal behaviour.

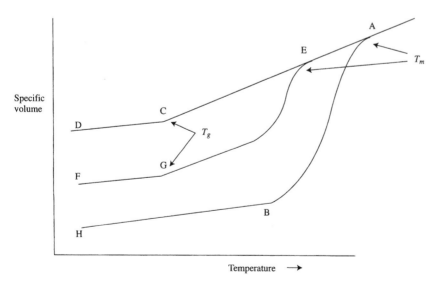

Fig. 2.21 Specific volume *versus* temperature (A, C, and D, amorphous; A, G, and F, semi-crystalline; A, B, and H, crystalline).

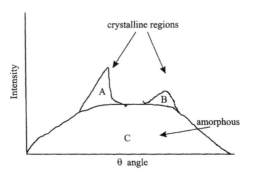

Fig. 2.22 X-ray data from a typical polymer showing crystalline/amorphous regions.

The second rule predicts that molten polymers crystallize at maximum rate at a temperature of 0.9 of the T_m (expressed in Kelvin). With polymers which are difficult to crystallize, it is convenient to be able to estimate the temperature at which the material ought best to crystallize. For example, isotactic polystyrene melts at 235 °C and the maximum rate of crystallization occurs at 184 °C. These types of empirical rules are widely used in polymer processing.

The degree of crystallinity may be measured by a number of techniques: X-ray diffraction, infrared spectroscopy or density measurements. X-ray is probably the most useful since the crystalline portions of the polymer produce sharp X-ray diffraction peaks while the amorphous region gives broad peaks (Fig. 2.22).

The degree of crystallinity is estimated from the relative areas under the two types of peak (i.e. percentage crystallinity is (A + B)/(A + B + C)).

Although the crystals melt over a temperature range, there is a temperature above which the crystalline form cannot exist. This is the melting point (T_m). It has been shown that the higher the number average molar mass ($\overline{M}_n$) the higher is the melting point for any given polymer. Intuitively the higher RMM would be expected to favour the solid state.

Relation of crystallinity and melting point to structure

Just as there are correlations between glass transitions and structure, there are also correlations between the degree of crystallinity and structure under the same broad headings: namely, chain flexibility, steric hindrance, side group effects and symmetry.

Chain flexibility

As with T_g this is most important for T_m. A comparison of polyethylene adipate with polyethylene terephthalate shows the stiffening effect of the backbone aromatic phenylene rings (Fig. 2.23) in increasing the melting point.

Qualitatively one can picture the effect of stiffness by comparing the peeling of a leather strap from a wall to which it is nailed, and the peeling of a wooden board from the wall. In the first case the nails are stressed sequentially. In the second case all the nails are stressed simultaneously. By analogy, more energy is required to separate molecules which are stiff than molecules that are flexible.

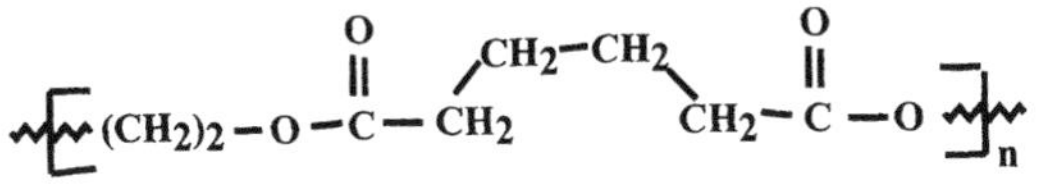

polyethylene adipate
$T_m = +50^0C$

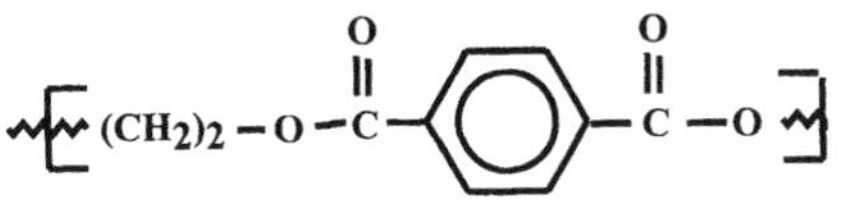

polyethylene terephthalate
$T_m = +267^0C$

Fig. 2.23 T_m's for polyethylenadipate/terephthalate.

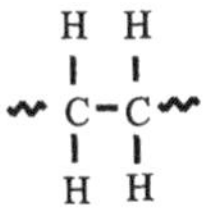

poly(ethene)
$T_m = +137^0C$

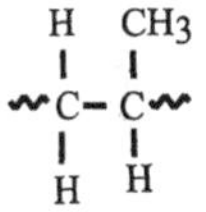

poly(propene)
$T_m = +176^0C$

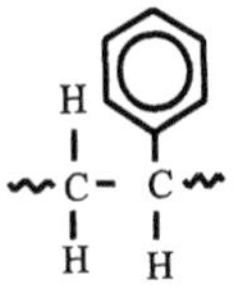

poly(styrene)
$T_m = +240^0C$

Fig. 2.24 Side group effects on T_m.

Quantitatively, since melting takes place without a change in Gibbs Free Energy (i.e. $\Delta G = 0$ in eqn 2.18).

$$\Delta G = \Delta H - T\Delta S = 0 \qquad (2.18)$$

then $T_m = \dfrac{\Delta H}{\Delta S} = \dfrac{H_L - H_C}{S_L - S_C}$ where H_L and H_C, and S_L are the enthalpies and entropies of the liquid (L) and the crystal (C) respectively. On melting, stiff rod-like chains undergo smaller entropy changes than do flexible chains. As the melting point is approached, the rods roll over one another without destroying the crystal lattice. This increased motion increases the entropy (the randomness) of the crystal, S_C, thereby lowering ΔS and hence raising T_m (eqn 2.18). Irregularly shaped molecules, i.e. flexible molecules, cannot roll without destroying the crystal lattice.

Side group effects
Using polyethene, polypropene and polystyrene as examples (Fig. 2.24), larger side groups increases T_m.

Higher T_m is representative of more crystallinity and better packing. What may be difficult to reconcile is that the introduction of a rigid phenyl ring has led to a higher melting point. In fact the rigid side groups produce repulsion and lead to a loss of planarity to give a helical structure having alternating trans and gauche positions along the polymer chain (Fig. 2.25).

However, the results in Fig. 2.24 refer to stereoregular polymers and it is important to recognize that the atactic polymer would prevent the formation of a regular helix. Clearly stereochemistry affects T_m and regular isotactic and syndiotactic polymers are more crystalline than random (atactic) ones. An increase in flexibility of side group leads to a lowering of T_m (Fig. 2.26); this is similar trend to the previously discussed effects on T_g.

Symmetry effects
Symmetrical molecules have higher melting points since they are more rod-like and are capable of rolling. For example, a para-substituted aromatic polymer would look exactly the same if the molecules were rotated through

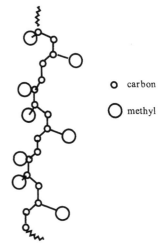

O carbon

O methyl

Fig. 2.25 Helix structure (rigid side groups produce repulsion and lead to a loss of planarity to give a helix. The helix form is regular and so the chains pack well to give crystals.

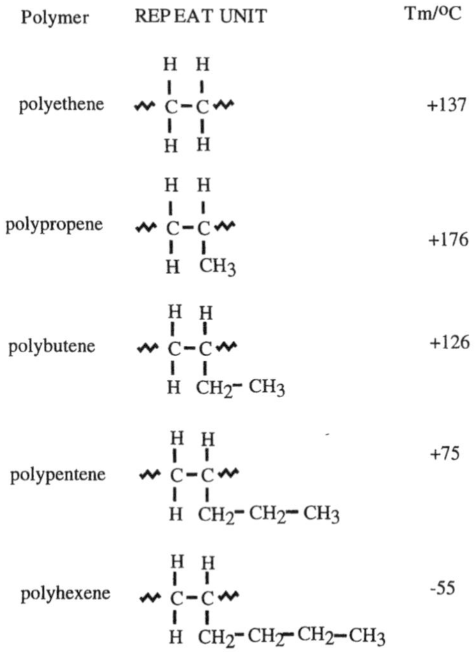

Polymer	REPEAT UNIT	Tm/°C
polyethene	$\sim$ C–C $\sim$ with H, H above and H, H below	+137
polypropene	$\sim$ C–C $\sim$ with H, H above and H, CH$_3$ below	+176
polybutene	$\sim$ C–C $\sim$ with H, H above and H, CH$_2$–CH$_3$ below	+126
polypentene	$\sim$ C–C $\sim$ with H, H above and H, CH$_2$–CH$_2$–CH$_3$ below	+75
polyhexene	$\sim$ C–C $\sim$ with H, H above and H, CH$_2$–CH$_2$–CH$_2$–CH$_3$ below	-55

Fig. 2.26 Effect of side chain flexibility

180°. However, the meta-linked analogue would have a different shape when reversed. Thus a meta compound can gain more entropy (randomness) by becoming free to move and tends to melt at a lower temperature (see Chapter 6, *High-performance Polymers*). This is the exact opposite to what was discussed for the T_g and is a fundamental difference between T_g and T_m.

Polarity

High cohesive energies and intermolecular forces (H bonds) tend to raise the melting point. Take for example the polyamides (i.e. nylons) (Fig. 2.27) where, as the distance between the amine and carboxylic acid groups present in the initial monomers increases (i.e. increase in n, Fig. 2.28), then there will be less H bonds in a given chain length and hence a lower T_m. N-methyl substituted nylons

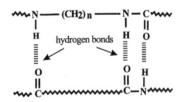

Fig. 2.27 Hydrogen bonding in nylons.

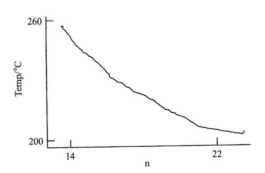

Fig. 2.28 Melting temperature *versus* n value for aliphatic polymides (nylons).

(i.e. $N - Me$ replaces $N - H$ in Fig. 2.27) melt at substantially (100–150 °C) lower temperatures than the parent nylon since they have no hydrogen bonding and also have less stabilization because of steric hindrance.

Monomer characteristics

One parameter which did not feature in the discussion of glass transition temperature was monomer melting point. If ethene ($T_m = -180$ °C) and acrylonitrile (propenonitrile $T_m = -82$ °C) are compared, it is found that the latter polymer (PAN, $T_m = 317$ °C) has a higher melting point than the former (PE, $T_m = 137$ °C).

Thus symmetry, flexibility, tacticity, and polarity alter T_g and T_m in the same sense, with only symmetry widening the gap between them. These parameters are important in applications such as clothes made from polymeric fibres. For example, it is preferable that a polymer has a $T_m > 200$ °C so that a garment can be ironed without melting, but needs a $T_m < 300$ °C to enable the fibre to be spun from the melt, whereas the T_g should lie between 30 and 120 °C, which allows for permanent creases. Obviously a cloth softens when ironed at about 200 °C so that any creases or pleats made will be retained on cooling. Subsequent washing is normally carried out a temperature too low to resoften the polymer significantly and so destroy the pleats. This permanent pleat is a desirable feature of some clothing and is deliberately sought.

2.4   Polymer processing

Polymer melts

The behaviour of polymer melts (rheology) is an important subject beyond the scope of this *Primer*, but it is important in controlling the choice of polymer processing conditions and therefore worthy of a brief description. In general, polymer melts (and also some solutions, particularly concentrated ones) can be non-Newtonian in behaviour; that is, the melt parameters are dependent upon shear rate or other stimuli. Although melt viscosity often decreases with shear force, it can increase if stress-induced crystallization occurs in the melt. Molar mass also affects melt behaviours since chain entanglements become more pronounced with greater chain length. Furthermore, although T_g is not pressure-sensitive, melt viscosity is significantly affected by pressure and this must be borne in mind when designing processing procedure. There is also a time-dependence of viscosity and a fluid whose viscosity varies with time under shear is called thixotropic. Many phenomena can also apply in reverse. For example, crystallization from the melt is time-dependent. Polymers do not simply crystallize as small-molecule compounds do; instead the chains fold over as sheets (lamellae) to give structures called spherulites (see Section 2.6.1).

General processing techniques

The convenient broad classification of polymers into thermoplastics, thermosets, and rubbers also defines their processing characteristics (Fig. 2.29). Although this is a wide subject, some general points may be made.

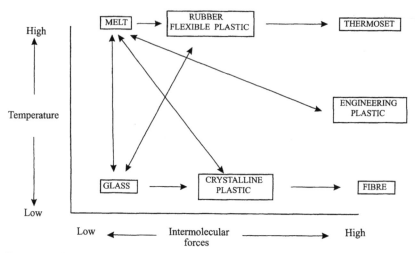

Fig. 2.29   States of bulk polymers.

Thermoplastics can be safely processed by melting and shaping the melt, after which cooling gives the final product. Scrap, such as reject mouldings, mould flash, etc., can be reprocessed by regrinding and repeating the process, although repeated processing may degrade the polymer chemically and special precautions in the form of antioxidants and heat stabilizing additives must be taken.

Thermosets will cure irreversibly during heating, usually because of chemical cross-linking (see Chapter 5). Here heating must be the last stage of processing and the process must be designed to ensure that shaping is complete before the irreversible chemical reaction has occurred. It is not usually necessary to cool the finished product before removal from the mould.

Rubbers are processed in the rubbery state and are then thermoset by vulcanizing. Rubber processing employs the rubbery state rather than the melt, i.e. the polymer is above its T_g and below T_m. Before vulcanization, the cross-links along the polymer chain are not present and the shaping processes are able to use viscous flow. Once vulcanization has taken place, viscous flow will no longer occur. This is a variation on the theme of thermosets.

Some branches of the polymer processing industries have evolved traditional ways of handling their materials, usually because they began by using polymers of natural origin. In most cases, these traditional methods have persisted into modern times because of the inertia of existing plant and skills. Particular examples are rubber, cellulose, and textiles. The rubber industry was founded on natural rubber which was obtained from trees as latex, dried and supplied as large, unwieldy bales. Powerful and specialized machinery was developed to handle the polymer in this form and today synthetic rubber is supplied in an identical type of bale. It would have been perfectly feasible to produce synthetic rubbers in other more easily handled forms, but the industry already had plant and skills for natural rubber and preferred its new materials in similar form. For example, in its early days PVC was largely processed by rubber manufacturers. This has resulted in the adoption of rubber processing techniques for PVC.

In the earliest days of polymer processing, techniques were evolved for moulding and casting cellulose-derived polymers and many of these persist today, e.g. film casting and solution spinning of fibres. Thus synthetic fibres are handled on a plant which is virtually identical to that devised for handling natural fibres such as wool and cotton, e.g. synthetic staple fibre is supplied in highly entangled compressed bales. The traditional processes of opening, carding, spinning, weaving, scouring, and finishing are still followed.

Although full discussion is beyond the remit of this *Primer* it is worth listing here the major typical processes used in fabricating polymer articles. These are extrusion, injection moulding, blow moulding, compression moulding, foam-cored moulding, reaction injection moulding (RIM), vacuum forming and calendering (roller-processing).

Additives

Since fabrication processes can take place at moderately high temperature and the resultant polymer product could be subject to a hostile in-service environment, the polymer is usually mixed with additives such as **process aids** (lubricants, antistatics), **stabilizers** (antioxidants, heat stabilizers, ultraviolet absorbers) and **modifiers** (plasticizers, vulcanizers, fillers, impact modifiers, blowing agents, and pigments) before fabrication. These additives may account for over 50% by mass of the resultant article. A full list of additives is extensive but mention should be made of the following.

Fillers
These are added to improve the properties and decrease the cost of the material. For example, finely powdered wood added to phenol-formaldehyde resins (Bakelite) makes it mouldable, decreases its brittleness and decreases the cost of the final product. Finely divided carbon added to rubber increases its tensile strength. Alumina and silica are cheap inorganic powders often added to bulk out polymers. Finely divided carbon is also added when electrical conductivity is desired, e.g. for antistatic applications such as car tyres (this is different to the intrisically conducting polymers discussed in chapter 6).

Pigments
Pigments are added solely for decorative purposes and should be able to withstand the high temperatures during processing. A feature of polymeric articles is that they can be coloured throughout the bulk material.

Cross-linking agents
These agents are used to produce cross-linkages between chains, thus turning a thermoplastic into a thermoset (see Chapter 5).

Blowing agent
These agents are used in polymer foams. They consist of chemicals which, when heated, release a gas which will 'blow up' the polymer while it is plastic. Bicarbonates have been used to release carbon dioxide but compounds such as azobisdiisobutyronitrile (AZBN which releases nitrogen) are now more commonplace.

Other process-enhancing additives include surfactants, lubricants, viscosity modifiers and mould-release agents ('sticky' polymers can have problems of adherence during processing).

Additives to counter environmental attack include antimicrobials, bacteriocides, fungicides, and similar species.

Flame retardants and smoke suppressants: since most polymers are inherently inflammable, this is an important aspect of polymer technology. Flames involve free-radical processes and so species which produce halogen atoms, particularly bromine atoms, are good flame retardants, e.g. hexabromocyclodecane, activated by antimony oxide. Such synergistic systems are effective, although there is increasing environmental concern over the presence of antimony (also arsenic salts) in some practical polymers.

Plasticizers

These are added to make processing easier or to make the final product softer or more flexible. Generally they act by interposing their molecules between the chains, so decreasing the attraction between the chains. Dinonyl and dioctyl phthalate are two liquids with boiling points above 200 °C which are used to make PVC more flexible. If the polymer is to be used under conditions in which a liquid plasticizer may be dissolved out of the material, plasticizers such as polypropylene adipate, which is itself a low molecular weight polymer, are used. Many resins with low RMM are similarly used as plasticizers. With the increase in convenience foods and use of microwave ovens, plastics for food require involatile plasticisers—see page 89.

Stabilizers

Most rubbers contain carbon–carbon double bonds. Some other polymers which one might not expect to contain such bonds also contain >C=C< bonds. For example, while PVC is normally written as shown in Fig. 2.30, a more accurate representation is Fig. 2.31. This imperfection is a result of side reactions. These double bonds are reactive and ozone (atmosphere) is capable of breaking the chains at these points, so degrading the material. Further, at 420 K, PVC breaks down forming double bonds with the HCl formed catalysing further reaction (Fig. 2.32).

Stabilizers are available to prevent these deleterious chemical changes, e.g. the addition of lead carbonate to remove HCl or antioxidants such as phenols which are oxidized more readily than the polymer. Many stabilizers are

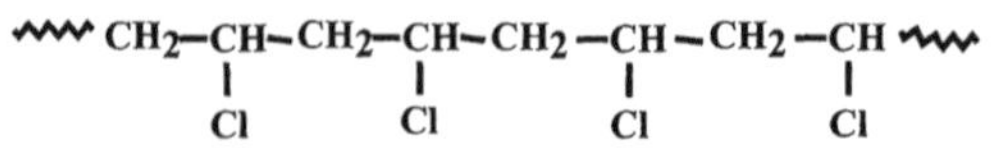

$$\text{\textasciitilde\textasciitilde\textasciitilde CH}_2\text{-CH-CH}_2\text{-CH-CH}_2\text{-CH-CH}_2\text{-CH \textasciitilde\textasciitilde\textasciitilde}$$
$$\qquad\quad | \qquad\quad | \qquad\quad | \qquad\quad |$$
$$\qquad\quad \text{Cl} \qquad\quad \text{Cl} \qquad\quad \text{Cl} \qquad\quad \text{Cl}$$

Fig. 2.30 Linear PVC.

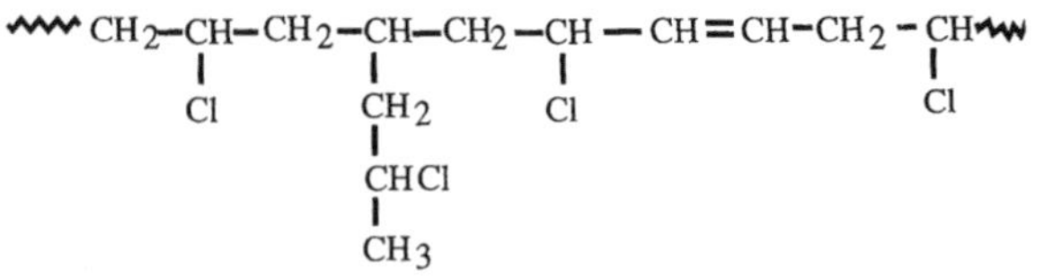

$$\text{\textasciitilde\textasciitilde\textasciitilde CH}_2\text{-CH-CH}_2\text{-CH-CH}_2\text{-CH — CH=CH-CH}_2\text{ - CH\textasciitilde\textasciitilde\textasciitilde}$$
$$\qquad\quad | \qquad\qquad | \qquad\qquad | \qquad\qquad\qquad\qquad |$$
$$\qquad\quad \text{Cl} \qquad\quad \text{CH}_2 \qquad\quad \text{Cl} \qquad\qquad\qquad \text{Cl}$$
$$\qquad\qquad\qquad\quad |$$
$$\qquad\qquad\qquad\quad \text{CHCl}$$
$$\qquad\qquad\qquad\quad |$$
$$\qquad\qquad\qquad\quad \text{CH}_3$$

Fig. 2.31 Branched PVC.

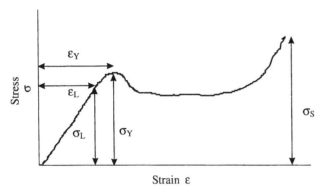

Fig. 2.32 Unsaturated PVC.

tailored towards a particular polymer and are intended to prevent processes to which it is susceptible.

In general, commercial polymers are surprisingly complex mixtures and this can affect strategies for recycling and environmental remediation (see Chapter 6, section 6.6).

2.5 Mechanical properties

Stress–strain measurements

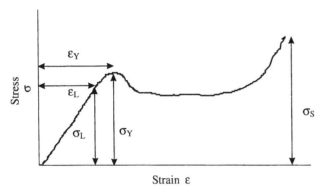

Fig. 2.33 Typical stress–strain curve.

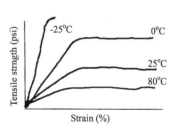

Fig. 2.34 Stress-strain curves for cellulose acetate at various temperatures.

Although these are the most widely used mechanical tests, unfortunately data tables are difficult to use because of variations in the properties of polymeric materials as a function of time, temperature, RMM, chain entanglement, crystallinity or other parameters. A typical stress–strain curve is given in Fig. 2.33.

Fig. 2.33 shows the important parameters of any material. For example, the tensile (Young's) modulus is defined as $E = d\sigma/d\varepsilon = \sigma_L/\varepsilon_L$; $\sigma_Y =$ Yield stress, $\varepsilon_Y =$ Elongation at yield and $\sigma_s =$ ultimate tensile strength (UTS). The acual shapes of the stress–strain curves found for polymeric materials depend on whether they can be regarded as soft and weak, or hard and brittle, etc. Unlike metals, polymers show a variation between these stress–strain behaviours as a function of temperature (Fig. 2.34).

What is obvious from Fig. 2.34 is that, as the temperature increases, then the modulus decreases, the yield strength decreases, the tensile strength decreases, and the curve changes from a **hard brittle** material (high modulus, small elongation to break) to a **hard tough** material (high modulus, large elongation) through to a **soft tough** material (low modulus, large elongation).

A plot of the viscoelastic modulus with respect to temperature is given in Fig. 2.35. It is possible to relate the various regions of the displacement curve (Fig. 2.35) to the deformation (or creep) characteristics of polymeric materials.

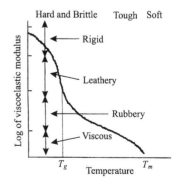

Fig. 2.35 Viscoelastic modulus *versus* temperature.

1. Below the T_g, where only **elastic deformation** can occur, the material is comparatively rigid, e.g. clear plastic triangle as used by draughtsmen.
2. In the region of T_g the material is **leathery**, i.e. it can be folded and deformed but does not spring back quickly to its original shape.
3. In the **rubber plateau** the material deforms readily, but quickly regains its previous shape if the stress is removed, e.g. a rubber ball or polythene 'squeeze' bottle as used in chemistry laboratories.
4. At very high temperatures (or under extensive loads) the material deforms extensively by **viscous flow** e.g. asphalt in a car park when a car is parked on a very hot day.

Obviously different structures will deform to different extents and Fig. 2.36 compares the behaviour of crystalline, amorphous, elastomeric and cross-linked polymeric materials.

Highly crystalline polymeric materials soften more gradually as the temperature increases until the melting point (T_m) is approached, at which point fluid flow becomes more significant. Vulcanized rubber is harder than unvulcanized rubber, and its curve would be raised more and more as a larger fraction of the possible cross-links are connected. It is important to recognize that the effects of cross-linking carry beyond the melting point into the true liquid. In this respect, a network polymer like phenol–formaldehyde is an extreme example of cross-linking. It gains its thermoset characteristics by the fact that the three-dimensional amorphous structure carries well beyond an imaginable melting point. For this reason it is used in rocket nose cones (networks are further discussed in Chapter 5).

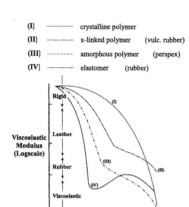

Fig. 2.36 Comparison of deformation behaviour for different structures.

Mechanical failure of polymers

Although softening and creep may be the first steps in terminating the usefulness of a polymeric material, failure by fracture, fatigue or tearing is of more concern to a consumer. For example, a plastic bag maybe strong enough to resist a tear — unless it already has a hole in it. Likewise, cracks in plastic toys usually occur at corners where the stresses are most concentrated. Thus, although the design engineer relies to some extent on stress–strain information, because of the problems associated with these measurements, especially in terms of temperature and the rate of application of the stress, they tend also to use special tests, e.g. bag-bursting, needle penetration, cloth tearing.

Fracture and tearing are called **ultimate failures** and are as much characteristics of dimensions and shapes of products as they are of geometry of load.

The actual mechanism of fracture is complicated since the structure of the polymer is often complex and the following is only a generality.

Below the T_g a material is brittle. This can lead to a fracture surface (at, say, a void, a crack, a notch or an imbedded particle) which cuts through the molecules with very little absorption of energy (Fig. 2.37).

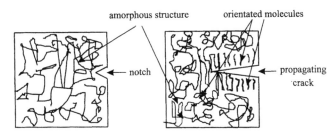
amorphous structure   orientated molecules

← notch

← propagating crack

Fig. 2.37 and 2.38 Material with notch and propagating crack.

Often, however, the propagating crack (Fig. 2.38) can cause local deformation of molecules with considerable consumption of energy. If energy can be absorbed by a reorientation of the molecules, the crack is slowed or even stopped. This type of behaviour is called **crazing** and the effect can sometimes be to **toughen** a polymer.

However, toughness is more usually built into materials by one of the methods below.

1. Addition of a filler. This could be particles of glass, or any material that interacts with the polymer matrix. Since the filler is usually harder, stronger, and more rigid than the polymer, any crack which develops must either remain in the polymeric phase or follow the phase boundary. If the filler can be made to bond strongly with the polymer matrix, the crack will stop moving. A typical glass reinforced plastic, 'fibreglass' (GRP) is ISOPON® (see Chapter 5).
2. Using polyblends of rubbery and glassy polymeric phases. The rubber section is capable of withstanding more strain, before failure, than rigid glassy section.
3. Make block copolymers with both crystalline and amorphous segments (see Section 3.3.4). The amorphous region gives an improved ability to withstand stress without failure, while the crystalline region improves ability to withstand large stresses.
4. Increase the amount of plastic flow available by adding a plasticizer. This pushes the molecules further apart and thereby increases flow and decreases brittleness. The disadvantage of this method is that yield stress will also decrease.

2.6 Other properties

Optical properties

Examples of where the transmission of light through a polymeric material is important are in the manufacture of contact lenses, wrappings, windscreens, etc. In general materials are either transparent (transmits all light), translucent (scatter some light) or opaque (scatter all light). Translucent and opaque materials scatter the light by reflection or refraction from the phase boundary between the polymer–additive or amorphous–crystalline regions.

The amount of refraction, n is given by c/v where v is the velocity of light in the material and c is the velocity of light in a vacuum. This phenomenon can be used in stress analysis.

For example, strains modify the index of refraction of any material. Light vibrating in one plane within a plastic which is under strain travels faster than light vibrating in a plane at right angles (i.e. not under strain). The difference between the two velocities, shows up as bifringence, $n_{slow} - n_{fast}$ and, with appropriate optical analysers, the distribution and areas of stress can be studied.

The phenomenon can also be used to identify the spherulitic regions in a material. (In simple terms, spherulites consist of numerous crystallite regions formed by nucleation at different points in the sample.) Here the structure of the material gives rise to destructive interference and with the help of polarized light the polymer scientist can investigate the structure. As an example of the property of a spherulitic material, examine what happens if the cellophane of a cigarette packet is folded in half several times. The material will appear initially transparent and will become increasingly more translucent (less transparent).

Electrical properties

The resistivity values of traditional polymers exceed 10^{12} Ohm cm compared to the values for metals which are approximately 10^{-5} Ohm cm, and compared to the value for semiconductors which are of the range 10^2–10^4 Ohm cm. Since resistivity is the inverse of conductivity, it is obvious that most polymers are insulators. However, one can make rubbers or plastics conductive by incorporating conducting fillers (typically graphite) into them. Thus, an appropriately compounded plastic or rubber can be used as the non-metallic ignition wires on cars or used to ground static charges.

Charge transportation or conductivity in plastics can also occur as a consequence of the transport of surface charges or surface conductivity. For example, polyesters contain ester linkages (Fig. 2.39) which provide exposed electrons on the oxygen atom adjacent to the surface. These polar groups can attract moisture (H_2O has a dipole), protons (H from butyric acid, a constituent of perspiration) or other surface contaminants (dirt is statically charged) and spoil the appearance of the article. To avoid this happening, one may use highly resistive non-polar coatings such as selected silicones.

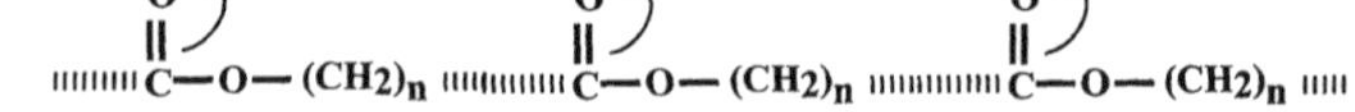

Fig. 2.39 Polyester charge distribution showing election density on the carbonyl oxygen.

There are new classes of functional polymer which have intrinsically high electrical conductivities. These are dealt with in Chapter 6.

Polymer Degradation

Chain scission

Polymers which have the longest molecules (hence biggest RMM) usually have the greatest strength and resistance to deformation (creep). This is due to the fact that the longer the molecule the greater the chain entanglement and therefore the more resistance to chain slippage and movement which leads to the permanent deformation. Obviously, at very high stresses the carbon–carbon backbone will rupture and break; this is the brittle nature of the polymer.

However, it is also possible to break a polymer backbone by treatment with high energy radiation, e.g. ultraviolet light, neutrons, or X-rays. Such a process maybe commercially important since it can lead to branching and cross-linking and hence higher molar masses, increased strength and resistance to creep and changes in the magnitude of T_g. In fact γ-rays have been used to modify the structure of low density polyethene (LDPE) and to produce highly branched material to be used as the insulating material for the wiring used in tunnels (see also Chapter 5 for electron-beam cross-linking).

An estimation of how much energy (E) is contained in one photon of ultraviolet light of wavelength (λ) 300 nm can be obtained using $E = hf$, where h is Planck's constant (6.62 $\times$ 10^{-34} Js) and f is the frequency of radiation (c/λ), c being the velocity of light (3 $\times$ 10^8 m/s). While the value obtained (6.62 $\times$ 10^{-19} J) may appear small, the bond dissociation energy of a carbon–chlorine bond, such as that found in PVC (polychloroethene), is 328 kJ/mol or 5.45 $\times$ 10^{-19} J/molecule. Since this value is less than the energy contained in a photon of ultraviolet light it is unsurprising that uPVC (unplasticized PVC) articles are not sold with long guarantees.

Thermal and chemical degradation

A particularly important class of degradation concerns chain-growth polymers, in which the reverse of chain build-up occurs. This can occur by scission in the middle of the chain, followed by break-up to oligomers, or by sequential loss of monomers from the end, in the exact reverse of propagation (i.e. unzipping). An example of the former is polystyrene, in which initial scission may occur at a chain defect, leading to proton transfer to give a short alkene-ended chain. Further heating produces a mixture of monomers, dimers, trimers, and beyond. An example of the latter is polymethylmethacrylate (PMMA).

There are other numerous specific chemical opportunities for degradation depending upon polymer chemistry. For example, step-growth polyester and polyamides can undergo main chain scission by hydrolysis, while side chain ester groups such as polyvinyl acetate and acrylates can also be hydrolysed. PVC can dehydrohalogenate. Polyurethanes are susceptible to bacterial attack, which involves environmental aspects of degradation (see Chapter 6). Allylic hydrogen atoms are susceptible to oxidation and these are found in natural rubber and dienes.

Not all degradation is negative. For example, polyacrylonitrile fibres (PAN) can be partly pyrolysed to give carbon fibre and the mechanochemical scission of polymer chains can also be useful in the control of RMM. Mechan-ochemical scission can conveniently be accomplished by the ultrasonic irradiation of a polymer solution. During irradiation, the chains cannot flex sufficiently quickly to follow the alternating sound pressure waves and as a result the bond breaks, giving radicals which can recombine with other fragments to give a final material with an average RMM rather than the extremes of RMM found in the original sample. Mechanochemical breakage of polymers to give radicals can be seen visually if an envelope is opened in a

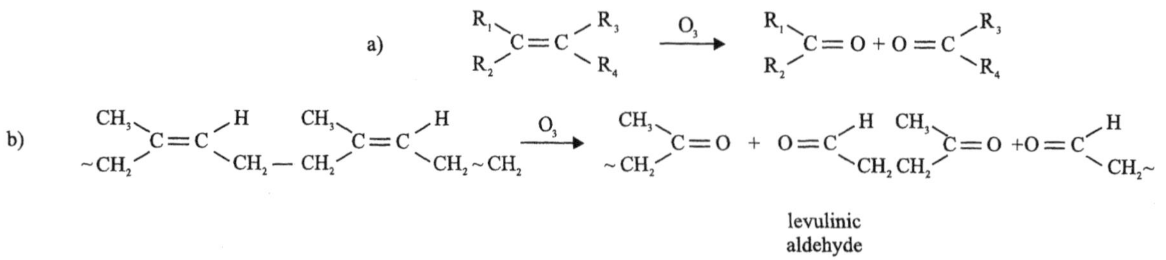

Fig. 2.40 Structural characterisation of natural rubber by ozonolysis
a) general principle
b) reaction of rubber

darkened room. A pale triboluminescence ensues from the breakage of bonds in the adhesive sticking down the flap of the envelope.

There are useful analytical studies which involve degradation. The classic example is the work on the ozonolysis of rubber by Harries in 1904. This reaction cleaves a double bond to give the two respective carbonyl compounds, through an addition–rearrangement–cleavage mechanism, Fig. 2.40a. Natural rubber gives laevulinic aldehyde which can be further oxidised to the carboxylic acid. This shows that rubber is composed of 1,4 linked isoprene units, as shown in Fig. 2.40b.

Swelling and permeation

Unlike chain scission, which is the breakage of strong intramolecular bonds, swelling is caused by the breaking of weaker intermolecular (chain to chain) bonds. Typically it is the contact of the material with solvent molecules which reduces the chain interactions and thereby causes the polymer chains to move apart, i.e. to swell, thereby becoming flexible and reducing the mechanical strength. Typical examples are the effects of water on polyvinyl alcohol and of petrol on rubber hose. The problem of swelling may be remedied by the following:

1. The use of highly cross-linked or crystalline polymers. Because of the presence of close meshed chains they will be resistant to swelling.
2. The use of incompatible polymer/solvent mixtures. For example, more polar neoprene is used as the petrol pipe in a motor car since petrol is non-polar, but non-polar isoprene is used as the brake fluid pipe, brake fluid being more polar.

The opposite to swelling is obviously shrinkage. Here the polymer–solvent molecular attractions are greater than either the solvent–solvent molecule attractions or the polymer–polymer chain attractions. The result is the chains are pulled closer together. A typical example is the effect of absorption of water into dry nylon.

Permeation is somewhat different to swelling and refers to the capability of small molecules to travel through a polymer matrix. It is very important in the protection of wrapped articles. Permeation of gases through polymers takes place by solution and diffusion, i.e. the permeating molecule must first dissolve in the outer layer of the film and then diffuse across it. Obviously the solubility depends upon compatibility in chemical structure. For example, water diffuses through cellophane a hundred times faster than through polythene, which is hydrophobic. Diffusion needs holes of comparable size to that of the permeating molecule, so the rate of diffusion falls off rapidly with increase RMM or as T_g is approached, i.e. rotation is hindered. The fact that water is capable of diffusion through cellophane, which is a semi-crystalline solid, is due to the fact that it acts as a plasticizer, thereby lowering the T_g and allowing rotation to occur.

2.7   Thermal methods of polymer analysis

This is a group of techniques in which some physical property of a substance is measured as a function of temperature or time while the substance is subject to a controlled temperature programme. The most common techniques are differential scanning calorimetry (DSC), thermal gravimetry (TG), dynamic mechanical analysis (DMA), dilatometry, and also applications-oriented specialized tests such as heat-deflection temperature and melt index.

Differential Scanning Calorimetry

Whenever a substance undergoes a phase change (or transition) such as melting (ice to water) or vaporization (water to steam), the temperature tends to remain constant (provided pressure is fixed) while energy is taken into the system to effect the phase transition, i.e. solid to liquid (latent heat of fusion) or liquid to gas (latent heat of vaporization). DSC is essentially a technique which compares the difference between the energy input into a substance and a reference (or blank) as a function of temperature (or time) while both the reference and the sample are subject to a controlled temperature rise (Fig. 2.41).

Since the blank contains no material and therefore cannot undergo any phase change, any energy input will simply raise the temperature. In order that the reference and sample are at the same temperature there is an accelerated inflow of energy to the sample. Fig. 2.42 shows a typical DSC of the polyester PET in which the T_g and T_m are clearly visible at 87°C (360° K) and 267°C (540° K) respectively.

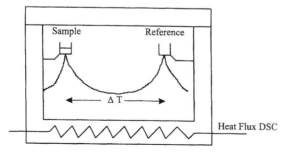

Fig. 2.41   Differential scanning calorimetry.

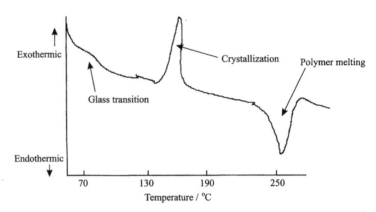

Fig. 2.42   Differential scanning calorimetry trace of PET.

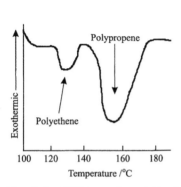

Fig. 2.43   DSC trace of a polyblend (polyethene–polypropene).

Also visible in the figure is the energy of crystallization. Once the material has sufficient energy for free rotation around the carbon–carbon bonds (i.e. above T_g), it is able to crystallize. Since a crystalline materials is more stable, this crystallization is accompanied by a release of energy, i.e. the converse of melting which required an energy input. DSC proves not only useful in characterizing a polymer by measuring the T_g and T_m, it can also be used to characterize polymer blends such as that between polyethene and polypropene (Fig. 2.43).

Thermogravimetry

In this method the change of mass of a polymer sample is measured as it is heated (Fig. 2.44). Weight loss by decomposition can be characteristic (Fig. 2.45 gives some typical curves) and can be performed under a range of gaseous atmospheres (air, nitrogen, etc.) to give further information while greater sophistication involves the use of mass spectrometry or some other means of trapping and identifying species that are volatized. If the identity of

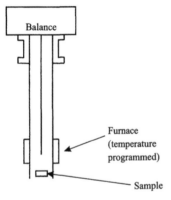

Fig. 2.44   Thermal gravimetry analysis instrument.

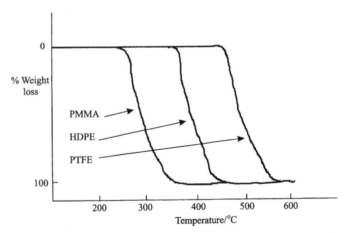

Fig. 2.45   Thermal gravimetry analysis trace (PMMA, polymethylmethacrylate; HDPE, high density polyethene; PTFE, polytetrafluoroethylene).

the species is important then other methods of volatilization such as secondary-ion mass spectrometry (SIMS) or laser-ion mass-analysis (LIMA) are useful, but these are not strictly thermal methods.

Dynamic mechanical analysis

In this technique the properties of a sample are studied as it goes through a time-dependent regime of mechanical change. This can give important information about relaxation processes and phase morphologies. There are considerable subtleties in the technique. The mechanical strain may be applied in a range of oscillatory manners, allowing resolution of the resulting stress into real and imaginary components, while the phase angle between stress and strain gives storage and loss moduli and loss factors. Low stress values are used to avoid permanent deformation. Experimental parameters include temperature, modulation frequency and amplitude. Modern apparatus can apply a variety of dynamic, compressive, tensional, and flexural strains. Subtle information can be obtained from the way that forces in the solid are dispersed. Some phase changes, difficult to detect by other methods, can be clearly seen by DMA. It is a powerful method used to identify segmental and side-chain motions within a chain and can be used on copolymers, additive-containing systems and other practical materials.

Dilatometry

This is an old technique but still useful. It involves measuring volume change with temperature. The sample is sealed into a bulb fitted with an accurate capillary and mercury (with its well known coefficient of thermal expansion) is used to fill the system. Upon gradual heating the volume expansion is measured as motion of the mercury level up the capillary. Deviations from a linear temperature–volume curve identify not only phase changes but also indicate their nature. A true first-order phase change such as melting is shown by a discontinuity in the trace (see Fig. 2.11), while a second-order phase change such as the T_g is shown by a change of gradient only.

Heat distortion (deflection) temperature

This is included to exemplify a practical applications-led method. Here reproducibility between different materials is obtained by following a specific protocol. The heat distortion temperature is defined by the (American) standard measurement test (ASTM) as the temperature at which a polymer bar of standard dimensions deflects by a fixed amount when heated with a standard load attached to its midpoint.

Melt index

This ASTM test refers to the mass in grams of molten polymer extruded through a standard size capillary at a known temperature for a known time. The values obtained depend on RMM and other properties of the sample. The test is used by manufacturers to compare the performance of materials in reproducible conditions relevant to practical applications. Many polymer tests are based on pragmatic measurement of practical behaviour rather than a quantification of the fundamental underlying parameters. Here extraction of rheological parameters would be a more challenging task.

3 Chain polymerization

3.1 Introduction

This class of polymers accounts for a large proportion of the synthetic polymer industry and includes the large-tonnage materials such as polyethene, polystyrene, polyvinylchloride (PVC) and acrylics. The mechanism of the reaction for all these is the opening of the π double bond in an alkene to give an all-carbon backbone of single σ bonds. (If a diene is used, then a double bond remains in the polymer, see later). However, since the empirical formula of the polymer produced (Fig. 1.11) is simply the sum of the requisite number of monomers, these are often called **addition polymers**, particularly in older texts. There is some ambiguity with step-type reactions such as polyurethane formation which does not evolve the loss of a small molecule (see Chapter 4).

$$nOH-R-OH + OCN-R'-NCO \rightarrow [O-R-OCO-NH-R'-NH-CO]_n$$
$$\text{Diol} \qquad\qquad \text{Diisocyanate} \qquad\qquad\qquad \text{Polyurethane}$$

The reaction mechanisms, which affect the build-up of polymer molar mass (RMM), and other factors are different for alkene polymers and polyurethanes. The alkene systems involve chain reaction mechanisms and the class of materials is better called **chain polymers**.

In general chain polymers can be prepared in one of three ways.

1. **Free-radical polymerization**, in which the alkene double bond opens homolytically (one electron in each direction). To do this an initiating species with an odd electron is required.
2. **Cationic polymerization**, in which an electron-deficient species removes both electrons from the electron-rich double bond, i.e. heterolytic fission produces a positively charged end-group to the propagating chain.
3. **Anionic polymerization**, in which a species more electron-rich than the double bond increases electron density of the π bonds such that heterolytic fission gives a negatively charged propagating chain end-group.

Table 3.1 shows the general trends in polymerization susceptibility for a number of alkene monomers. (Co-ordination complex catalysts ['Ziegler Natta' systems], which are anionic surface-active systems are not included in the table).

In general chain polymerizations have the following common features:

1. An initiation step in which a reactive species is generated and attacks the first monomer molecule.
2. A propagation step in which a large number of further monomers are sequentially added to give the long polymer chain, still retaining the reactive end-group.

Table 3.1 Susceptibility of monomers to type of initiation for chain polymerization

Monomer	Structure	Initiation method		
		Cationic	Free radical	Anionic
Isobutene	$H_2C = C(CH_3)_2$	✓	✗	✗
Vinyl ethers	$H_2C = CHOR$	✓	✗	✗
Ethylene	$H_2C = CH_2$	✓	✓	✗
Vinyl esters	$H_2C = CHOCOR$	✗	✓	✗
Vinyl halides	$H_2C = CHX$	✗	✓	✗
Acrylic esters	$H_2C = CHCOOR$	✗	✓	✓
Acrylonitrile	$H_2C = CHCN$	✗	✓	✓
Vinylidene halides	$H_2C = CX_2$	✗	✓	✓
Styrene	$H_2C = CHC_6H_5$	✓	✓	✓

R is a representative alkyl group and X is a halogen

3. A termination step in which the reactive end-group is deactivated. (This must eventually happen or the polymer would grow indefinitely.)

These features each have different kinetic regimes and control of the various stages of the reactions is of fundamental importance in commercial chain polymer synthesis. A number of strategies have been developed to produce materials of controlled RMM and properties.

3.2 Free radical vinyl polymerization

A free radical is a species containing a spare unpaired electron. It is highly reactive and undergoes reactions to extract an electron from another substrate to produce a full complement of electrons. Consider the example of the reaction of a methyl radical with hydrogen (eqn 3.1):

$$CH_3 \bullet + H{-}H \rightarrow CH_4 + H\bullet \tag{3.1}$$

Here the methyl radical has caused the σ bond of the hydrogen to be broken homolytically and has formed its own σ bond between the C and H, leaving another free radical H•. This free radical is now the active species and will attempt to undergo reaction with another species to produce a full complement of electrons.

Activation (termed 'initiation' for polymers) is produced by splitting an initiator molecule (X_2) homolytically, usually by thermal degradation (eqn 3.2), and allowing the odd-electron species to react with the monomer, here $CH_2 = CH_2$, by opening the double bond (eqn 3.3):

$$X{-}X \rightarrow X\bullet + \bullet X \tag{3.2}$$

$$X\bullet + \underset{\substack{|\\H}}{\overset{\substack{H\\|}}{C}} = \underset{\substack{|\\H}}{\overset{\substack{H\\|}}{C}} \rightarrow X - \underset{\substack{|\\H}}{\overset{\substack{H\\|}}{C}} - \underset{\substack{|\\H}}{\overset{\substack{H\\|}}{C}} \bullet \tag{3.3}$$

In general terms this can be written as shown in eqns 3.4a and b:

$$I \rightarrow 2R\bullet \tag{3.4a}$$

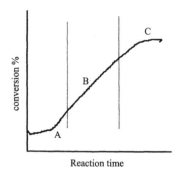

conversion %

Reaction time

Fig. 3.1 Conversion time curve for a chain polymerization. (A) During this phase the rate increases because of the build-up of the radical concentration in the system and the removal of impurities by the radicals (eqn 3.4a); (B) here there is a constant rate of polymerization (eqns 3.5a-c); (C) the final stage represents the mass action effect as the monomer is depleted and the radical species are terminated (eqn 3.6).

$$R\bullet + M \rightarrow RM\bullet \tag{3.4b}$$

The activated monomer (RM•) is now free to react with more monomers (eqns 3.5a–c).

$$RM\bullet + M \rightarrow RMM\bullet \quad (RM_2\bullet) \tag{3.5a}$$

$$RMM\bullet + M \rightarrow RMMM\bullet \quad (RM_3\bullet) \tag{3.5b}$$

$$RM_{x-1}\bullet + M \rightarrow RM_x\bullet \quad (RM_x\bullet) \tag{3.5c}$$

Deactivation of the growth process is called termination and occurs when two of these radical species combine to produce a 'dead' polymer as follows (eqn 3.6):

$$RM_x\bullet + RM_x\bullet \rightarrow RM_x - M_xR \tag{3.6}$$

A typical conversion graph for the initiation of a monomer M by hydrogen peroxide is given in Fig. 3.1.

If A and B were the only reactions, then the rate would continually increase towards explosion. Reaction B neither produces nor destroys radicals, so a balance must occur between A and C giving a fixed rate of polymerization i.e. section B on the graph shows a constant rate. B is known as a 'stationary state' where the rates of production (A) and destruction (C) are equal. This is a useful concept in polymer kinetics since the assumption that the rate of production of the active species is equal to the rate of termination allows an estimation of the concentration of active species.

Free radical initiators

Certain monomers have sufficiently reactive double bonds to undergo polymerization on heating (i.e. styrene, methyl methacrylate). Most others require an initiator. There are a large number of initiators available. Some of the most common types follow.

Peroxides (R–O–O–R) and hydroperoxides (R–O–O–H)
The most common peroxide is benzoyl (diaroyl) peroxide (Fig. 3.2a) which decomposes on heating.

Further reactions can occur, either to produce additional radical species (Fig. 3.2b), or else self-termination (Fig. 3.2c).

Other common peroxides are acetyl (ethanoyl) peroxide ($CH_3COOOCOCH_3$), the hydroperoxides ($ROOH \rightarrow RO\bullet + \bullet OH$) and the inorganic peroxide potassium persulphate. Ethanoyl peroxide provides radicals soluble in organic media, while the hydroperoxides provide both water-soluble hydroxyl radicals ($\bullet OH$) and radicals ($RO\bullet$) which are soluble in organic media.

Ideally an initiator should be relatively stable at room temperature but decompose rapidly at the processing temperature. For example, the activation energy for the breakdown of benzoyl peroxide is approximately 120 kJ/mol, making the decomposition fairly rapid at 50 °C. At 100 °C benzoyl peroxide has a half-life of 30 minutes, which has the advantage that benzoyloxy radicals are stable enough to react with the monomer before eliminating carbon dioxide. In contrast, acetoxy radicals ($CH_3COO\bullet$) are less stable and this leads to initiator wastage.

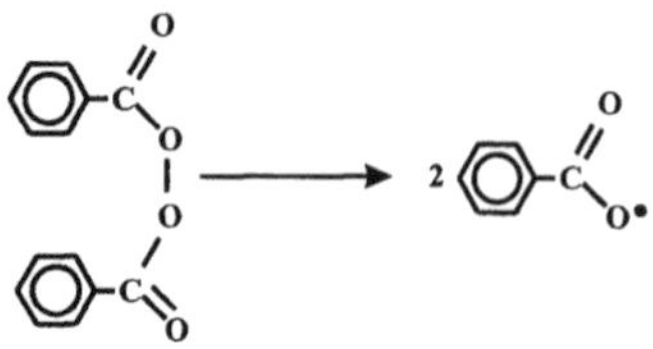

(a)

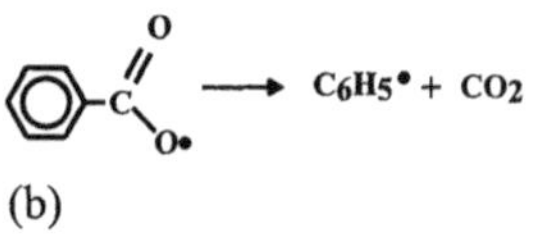

(b)

$$C_6H_5\bullet + CO_2$$

(c)

$$2C_6H_5\bullet \rightarrow C_6H_5\text{-}C_6H_5 \quad \text{(diphenyl)}$$

Fig. 3.2 Reactions during decomposition of benzoyl peroxide (free radical initiator).

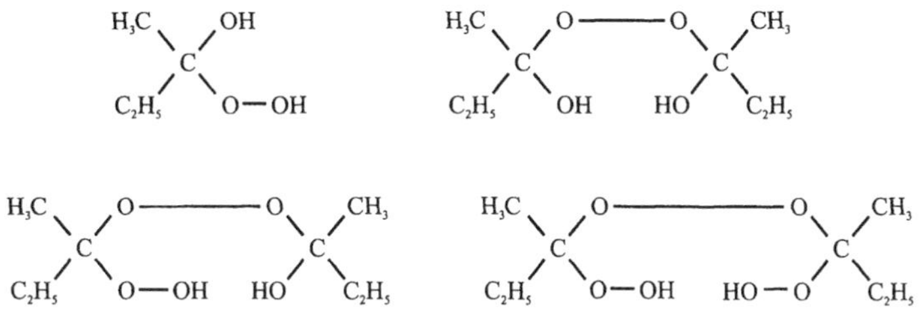

Fig. 3.3 'Methyl ethyl ketone peroxide': representative species formed from aerial oxidation of methyl ethyl ketone.

Industry seeks economy and a useful peroxide system is made by aerial oxidation of cheap methyl ethyl ketone to give a range of peroxy and hydroperoxy species which are used directly onwards (Fig. 3.3).

Peroxide decomposition can also be induced to occur at *lower* temperature by the addition of **promoters** e.g. the addition of N,N'-dimethylaniline to benzoyl peroxide. Redox transition metal salts, with variable valency such as iron, manganese or cobalt salts are also effective (Fig. 3.4).

Single electron transfer is a useful means of initiating low temperature polymerization or emulsion polymerization. For example, the decomposition of cumyl hydroperoxide is used in the emulsion polymerization of styrene and butadiene in forming SB rubber (eqn 3.7).

$$Co^{2+} + ROOH \longrightarrow RO\cdot + OH^{\ominus} + Co^{3+}$$

$$Co^{3+} + ROOH \longrightarrow ROO\cdot + H^{\oplus} + Co^{2+}$$

Fig. 3.4 Transition metal ion as accelerator for peroxide decomposition.

$$C_6H_5 - C(CH_3)_2 - O - O - H + Fe^{2+} \rightarrow C_6H_5 - C(CH_3)_2 - O \bullet + OH^- + Fe^{3+}$$

$$(3.7)$$

An example of an inorganic peroxide is potassium persulphate, $K_2S_2O_8$ (eqn 3.8, shown being activated by thiosulphate reducing agent).

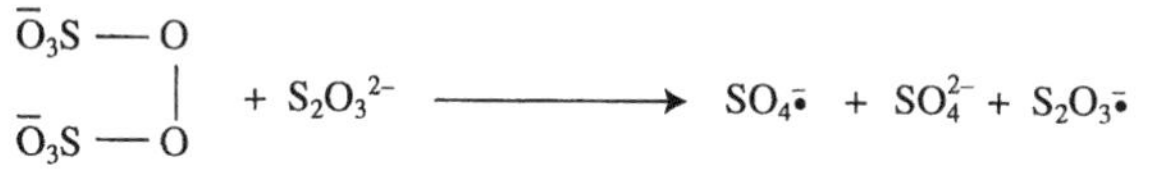

$$\begin{array}{c} \bar{O}_3S - O \\ | \\ \bar{O}_3S - O \end{array} + S_2O_3^{2-} \longrightarrow SO_4\bar{\cdot} + SO_4^{2-} + S_2O_3\bar{\cdot} \qquad (3.8)$$

Ferrous ions can also catalyse persulphate decomposition to radicals.

Azo compounds
The most common is azobisisobutyronitrile (AIBN or AZBN). It decomposes at relatively low temperatures and has a half-life of 1.3 h at 80 °C. It can also be induced to decompose at room temperature by irradiating with ultraviolet light ($\lambda = 360$ nm). The decomposition reaction is represented by eqn. 3.9.

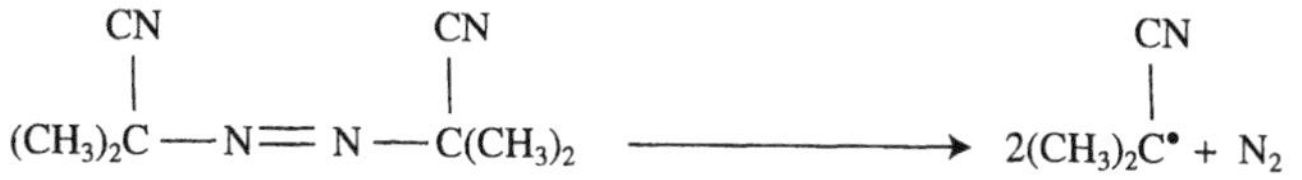

$$\begin{array}{ccc} CN & CN & CN \\ | & | & | \\ (CH_3)_2C - N = N - C(CH_3)_2 & \longrightarrow & 2(CH_3)_2C^\bullet + N_2 \end{array} \qquad (3.9)$$

This is an extremely important initiator which gives a clean reaction system and is useful in academic research. All free radical polymerization is known to be inhibited by the presence of oxygen, which itself is a diradical. However, the use of AIBN produces nitrogen as the reaction product and guarantees an inert atmosphere.

Photoinitiators

A major advantage of using photolysis is that it is independent of temperature thereby enabling polymerization to occur at low temperature and with better control since narrow wavelength bands may be used. Examples of other typical photoinitiators are disulphide (eqn 3.10), benzoin (eqn 3.11), benzil (eqn 3.12) as well as other peroxides and azo species similar to AZBN.

$$\underset{\text{disulphide}}{\text{RSSR}} \quad \overset{h\nu}{\to} \quad 2\text{RS}\bullet \tag{3.10}$$

$$C_6H_5 - \overset{\overset{O}{\|}}{C} - \overset{\overset{OH}{|}}{\underset{\underset{H}{|}}{C}} - C_6H_5 \quad \xrightarrow{h\nu} \quad C_6H_5 - CO\bullet + \bullet\overset{\overset{OH}{|}}{\underset{\underset{H}{|}}{C}} - C_6H_5 \tag{3.11}$$

benzoin

$$C_6H_5 - \overset{\overset{O}{\|}}{C} - \overset{\overset{O}{\|}}{C} - C_6H_5 \quad \xrightarrow{h\nu} \quad 2C_6H_5 - CO\bullet \tag{3.12}$$

benzil

Kinetics of free radical polymerization

The rate of any chemical process is proportional to the concentration of the reactant. Thus, as the reactant is used up the rate of the reaction will decrease. This can be seen by considering a barrel, initially full of water, draining through a hole at the base. Initially the pressure of water will cause the barrel to drain fairly quickly. However, as the water level falls, so does the pressure and the barrel will not drain as quickly. Thus in terms of kinetics we can represent the rate of reaction between A and B to produce C as:

$$A + B \overset{k}{\to} C \tag{3.13}$$

From simple kinetics, the rate of the reaction in terms of the loss of reactant A (and loss of B and gain of C) may be given as eqn. 3.14.

$$\left(\frac{-d[A]}{dt}\right) = k[A][B] = \left(\frac{-d[B]}{dt}\right) \text{ or } \left(\frac{+d[C]}{dt}\right) \tag{3.14}$$

The remainder of this section will use the ideas of simple kinetics to show how the rate of polymerization and RMM of the polymer depends upon the concentration of initiator and monomer. Throughout the section the initiator molecule, free radical and monomer are referred to as I, R and M respectively. The rates of the various processes, i.e. initiation, propagation and termination are given the symbol R_i, R_p and R_t and the various rate constants of initiation, propagation, etc. are given the symbols k_i, k_p etc.

Initiation

We have previously seen that the slow step in the initiation is given by eqn. 3.4a. Since the production of the active centres (R$\bullet$) occurs twice as fast as the loss of the initiator molecule ($-d[I]/dt$) and not all the radicals will be involved in activating the monomer since some radicals may undergo recombination (see diphenyl production, Fig. 3.2c), the production of active centres is given by eqn. 3.15.

$$\left(\frac{+d[R\bullet]}{dt}\right) = 2fk_i[I] \qquad (3.15)$$

where f represents initiator efficiency ($0 < f < 1$)

Propagation
Once the monomer has been activated there follows the rapid addition of more monomer units, e.g. $RM\bullet + M \rightarrow RMM\bullet \rightarrow RMMM\bullet$, etc. In the case of ethene, a symmetrical monomer, there is no preference for which carbon atom is attacked. However, with chloroethene, where one of the hydrogen atoms has been replaced by a chlorine atom, the left and right carbon atoms are no longer identical (Fig. 3.5).

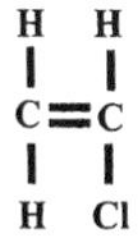

Fig. 3.5 Vinyl chloride.

When vinyl chloride is polymerized into chains, there are two ways two molecules can combine together, i.e. 'head to tail' (Fig. 3.6), which is the usual way, or 'tail to tail' (Fig. 3.7).

In practice, head to tail is by far the most common occurrence, certainly for a monomer ($CH_2=CHR$) containing a bulky R constituent. This is because the bulky R group prefers an unhindered methylene (CH_2) group on either side. Head to tail is also favoured where the monomer possesses a dipole, since the monomers line up the same way round to minimize the charge interaction.

No matter what type of addition takes place, all the monomer consumed eventually finds itself in the polymer. Thus the rate of loss of monomer ($-d[M]/dt$) can be equated to the propagation rate (R_p), or the rate of production of polymer ($+d[P]/dt$). In its simplest form, the rate of propagation (polymerization), R_p can be given by eqn. 3.16.

$$-d[M]/dt = k_p[R\bullet][M] \qquad (3.16)$$

Termination
Termination (or deactivation) occurs when any two radical species combine to produce 'dead' polymer (eqn 3.6). While in general there are two forms of

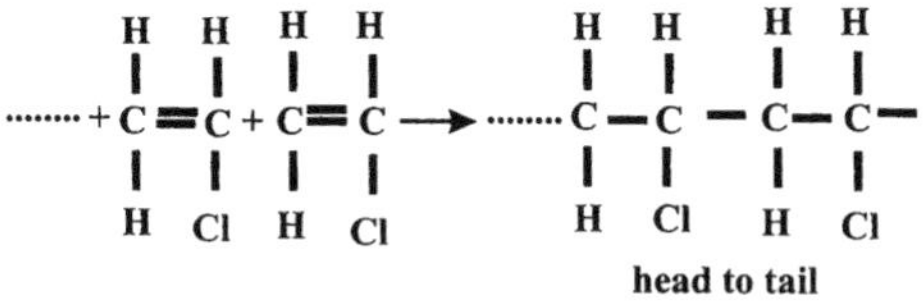

Fig. 3.6 Polymerization of vinyl chloride head to tail.

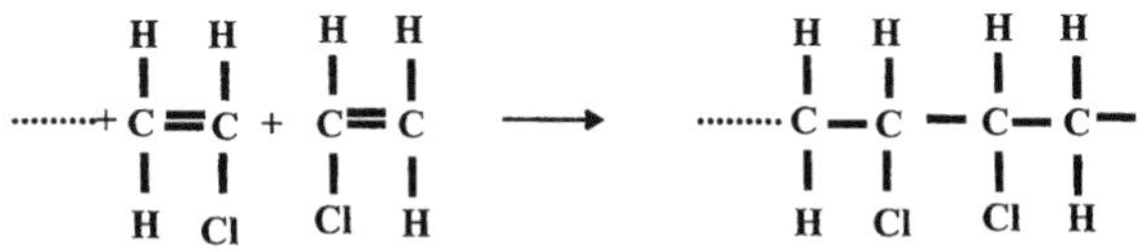

Fig. 3.7 Polymerization of vinyl chloride tail to tail.

termination, namely combination and disproportionation, the coupling of two radicals together (combination) is by far the less energetic, requiring approximately 20 kJ mol. Here termination will be restricted to the combination reaction shown in eqn. 3.6 so that the kinetic equation can be represented by eqn. 3.17.

$$R_t = 2k_t[R\bullet]^2 \tag{3.17}$$

(The factor of 2 comes from the fact that two chains are terminated in such a coupling step.) Invoking the Steady State Hypothesis, i.e. the rate of initiation equals the rate of termination (or $R_i = R_t$) and assuming f = 1 allows the derivation of eqn 3.18.

$$R_p = (k_p)\left(\frac{k_i}{k_t}\right)^{\frac{1}{2}}[M][I]^{\frac{1}{2}}$$

$$\text{or} \quad R_p = k'[M][I]^{\frac{1}{2}} \tag{3.18}$$

where $k' = (k_p)\left(\frac{k_i}{k}\right)^{\frac{1}{2}}$

Degree of polymerization
This is given the symbol X_n or D_p (in older texts) and is defined as the number of repeat units in a polymer chain. As such it is clearly related to chain length and the polymer's molar mass $\overline{M_n}$. For example $X_n = \dfrac{\overline{M_n}}{M_o}$ where $\overline{M_n}$ is equal to the number average RMM and M_o is the monomer RMM. Since the propagation step (R_p) is the step which consumes monomer, and the termination step (R_t) is the step which leads to the production of polymer chains, then the number of units in a chain is given by eqn. 3.19.

$$X_n = \frac{R_p}{R_t} = \quad \cdot \quad \frac{k_p[M]}{2k_t\left(\dfrac{k_i}{k_t}\right)^{\frac{1}{2}}[I]^{\frac{1}{2}}}$$

$$\text{or} \quad X_n = \frac{k''[M]}{[I]^{\frac{1}{2}}} \tag{3.19}$$

In summary, we now have two equations which characterize the effect of both monomer and initiator concentration on the polymerization rate and resultant molar mass.

$$R_p = k'[M][I]^{\frac{1}{2}} \tag{3.20}$$

$$X_n = \frac{k''[M]}{[I]^{\frac{1}{2}}} \tag{3.21}$$

From these equations the following may be deduced:

1. Rate will increase as both [M] and [I] increase.
2. RMM will decrease as [I] increases. Varying both [M] and [I] is a means of controlling RMM.

3. An increase in T increases k_i, k_p and k_t. In practice it is observed that the overall rate of conversion is approximately two to three-fold per 10 °C rise in temperature. Since RMM is inversely proportional to k_i and k_t, it is reduced as the temperature is raised.

In all previous kinetic discussions it has been assumed that termination was by coupling:

i.e. $R-(CH_2)_mCH_2\bullet + \bullet CH_2(CH_2)_n-R \rightarrow R-CH_2(CH_2)_{n+m}CH_2-R$

However, termination may occur also by **disproportionation** in which a hydrogen atom is abstracted or 'transferred' from the carbon atom 'next-door' to the site of the radical (Fig. 3.8). This produces one deactivated chain and one which still has a double bond present and can be re-initiated.

However, the energy requirements are the same whether the H is abstracted from active radical or from a dead polymer (Fig. 3.9). The mechanism given in Fig. 3.9 assumes that the monomer is $CH_2=CHY$ and shows that the process involves *inter*molecular H abstraction. This generates a new radical site on a polymer chain. This new site is able to grow, giving chain branching. This can affect a polymer's ability to crystallize and hence its mechanical strength.

This abstraction process may also occur *intra*molecularly to produce pendant side-chains and is commonly called 'backbiting', as shown in Fig. 3.10. It is most common with the unsubstituted alkene ethene where there is no R group for steric hindrance. It explains why early routes to polyethene gave a highly branched material with poor physical properties.

This abstraction process, or chain transfer as it is known, may also occur with the initiator, monomer or solvent. For example, polystyrene prepared in CCl_4 contains chlorine at the chain ends as a result of chain transfer as shown in Fig. 3.11.

$R - (CH_2)_mCH_2 \bullet + \bullet CH_2(CH_2)_n - R$
$\rightarrow R - CH_2(CH_2)_nCH_3 +$
$CH_2 = CH - (CH_2)_{m-1} - R$

Fig. 3.8 Termination by disproportionation.

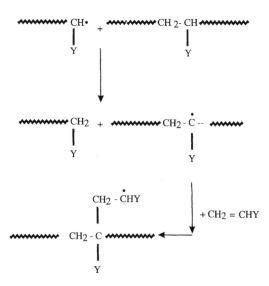

Fig. 3.9 Abstraction from –CHY.

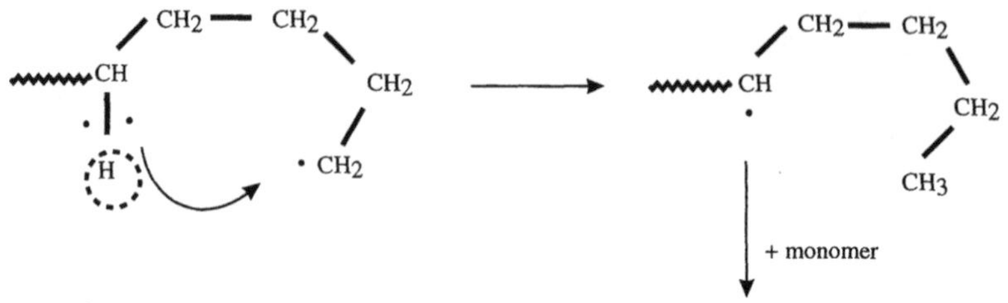

Fig. 3.10 Backbiting.

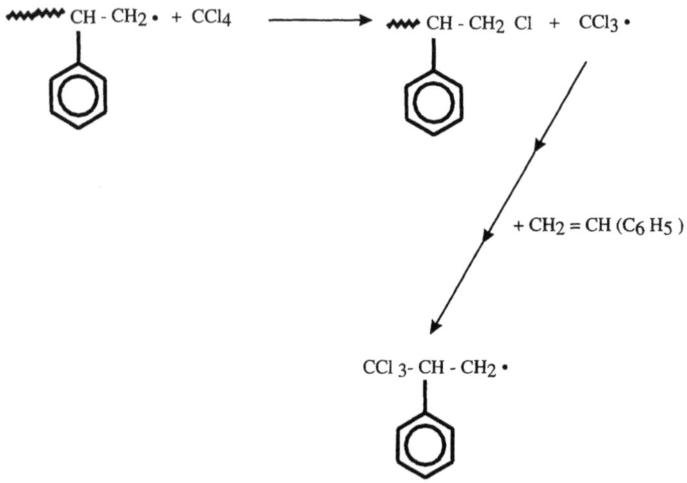

Fig. 3.11 Chain transfer of polystyrene prepared in carbon tetrachloride.

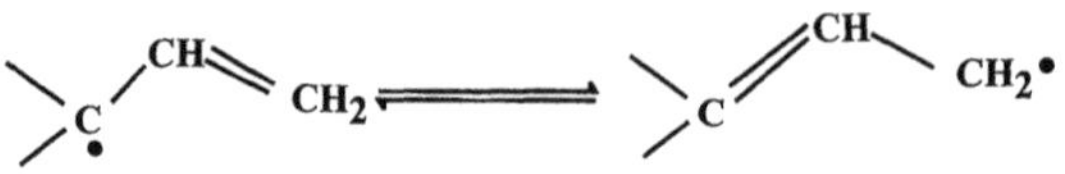

Fig. 3.12 Allylic hydrogen.

Transfer to monomer is important with monomers containing allylic hydrogens, i.e. CH_2–CH=CH_2 since the allylic radical is resonance-stabilized. For this reason high molecular weight (i.e. high molar mass) poly(propene) cannot be prepared by radical polymerization (Fig. 3.12).

Chain transfer is very important in the control of RMM and thus also important in determining polymer properties. We can define a **chain transfer agent** as a substance which is added to a polymer recipe to control RMM. These agents have a high affinity for H transfer. An example is a mercaptan, RSH.

$$\text{----}\ CH_2\text{-}\underset{\displaystyle Y}{\underset{|}{CH}}\bullet + RSH \longrightarrow \text{----}\ CH_2\text{-}\underset{\displaystyle Y}{\underset{|}{CH_2}} + RS\bullet$$

Obviously the transfer agent cannot recognize the difference between RM•, RMM•, RMMM• or RM$_n$•, so the global value R• is used. Denoting the general reaction of the radical, R•, with the transfer agent, T, as

$$R• + T \rightarrow R - H + T•$$

the kinetic equation can be written as eqn 3.22.

$$R_{tr} = k_{tr}[R•][T] \tag{3.22}$$

Since there are now two processes which can terminate the radical growth and produce polymer, i.e. termination and transfer, the degree of polymerization, X_n is redefined. In such cases the degree of polymerization is as eqn. 3.23.

$$(X_n)_{tr} = \frac{R_p}{R_t + R_{tr}} \tag{3.23}$$

Since $\frac{R_p}{R_t}$ is equal to X_n, the degree of polymerization in the absence of transfer, then eqn 3.23 becomes eqn 3.24.

$$\frac{1}{(X_n)_{tr}} = \frac{1}{X_n} + C_T \frac{[T]}{[M]} \tag{3.24}$$

where $C_T = \frac{k_{tr}}{k_p}$, the **transfer constant**. Therefore, as the transfer constant and/or the concentration of transfer agent increase, the degree of polymerization and hence RMM get smaller. This means that if the solvent can act as transfer agent, it can affect the build-up of polymer RMM. Polystyrene produced in CCl$_4$ has a significantly lower RMM than when produced in the less reactive benzene solvent. (Table 3.2)

Chain transfer reactions occur in most free radical vinyl polymerizations and relatively broad molar mass distributions are common.

Table 3.2 Chain transfer constants C_T for free radical polymerizations

Transfer agent	Styrene	Methyl methacrylate
Benzene	0.023	0.040
Cyclohexane	0.031	0.10
Chloroform	0.5	0.2
CCl$_4$	90	2.40
CBr$_4$	22 000	2 700
n-BuSH	210 000	6 600

Inhibitors and retarders

Certain substances react with free radicals to produce species which are incapable of initiating polymerization. If this process is so efficient that polymerization is prevented, the substances are known as **inhibitors**. If the substance merely slows the polymerization reaction it is called a **retarder**. Oxygen is both a good retarder and inhibitor for free radical reactions and must be excluded by working in an inert atmosphere.

Normally inhibitors are added to monomers in low concentration to prevent polymerization during transportation and storage. Therefore it is necessary to remove the inhibitor from stored monomer before use. On an industrial scale this may not be economical and either an inhibitor is used which is inactive in the presence of oxygen or an increased level of initiator is used to overcome the inhibitor before polymerization can occur.

3.3   Ionic chain polymerization

General aspects

Radical polymerization is non-selective with termination occurring by combination or disproportionation. In ionic polymerization the growing centre can be either negatively charged (anionic) or positively charged (cationic). Therefore there is little likelihood in ionic polymerization of termination by combination since like charges repel.

Although most alkenes will be initiated by any free radical, initiation by anion (or cation) will depend upon the substituents on the alkene, i.e., the inductive or resonance characteristics of the substituent (see Table 3.1 on page 51). For example electron donating substituents (G) such as alkoxy (OR), alkyl (R) or phenyl (C_6H_5) increase the electron cloud density on the C=C bond and further facilitate bonding to cationic species as shown in Fig. 3.13.

Substituents can also stabilize the new charge by resonance as shown in Fig. 3.14.

On the other hand the electron-withdrawing cyano (CN) and carbonyl (CO) groups stabilize an incipient anion as shown in Fig. 3.15 (see Table 3.1).

As mentioned previously, radical species are neutral and bring about the polymerization of almost all C=C bonds irrespective of polarity of the solvent. However, since ionic polymerization is by charged species, the type of solvent is important. High dielectric solvents allow complete separation of the ions. Low dielectric solvents keep ions together as ion pairs. Free ions, being less sterically hindered, are able to propagate (react) more than 1000 times faster than ion pairs (Fig. 3.16).

Although there is no combination in ionic polymerization, there may be charge transfer to solvent. This is highest for the highest dielectric medium, giving rise to fast polymerization but low molar mass products.

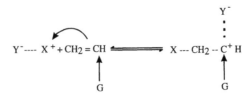

Fig. 3.13 Stabilization of cationic species by electron donating group (G).

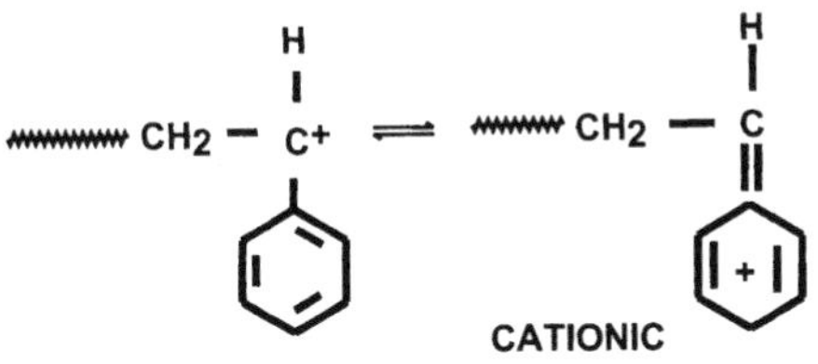

Fig. 3.14 Stabilization of charge by aromatic ring.

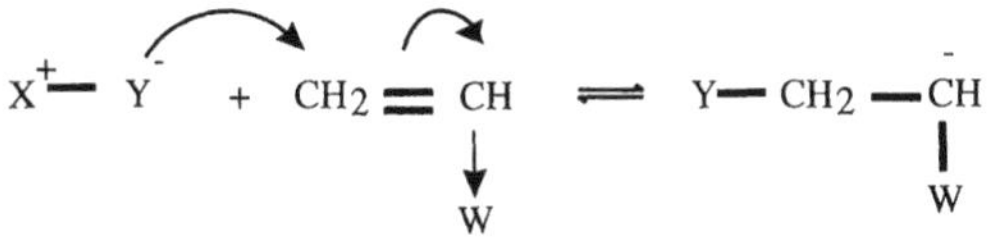

Fig. 3.15 Stabilization of anion by electron-withdrawing group (W).

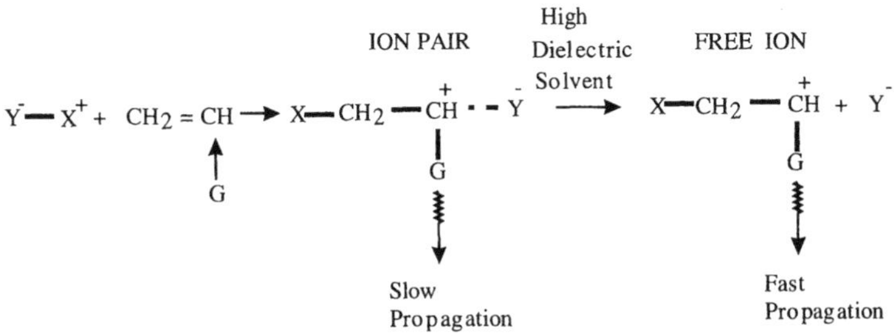

Fig. 3.16 Free ions versus ion pairs.

3.3.2 Cationic polymerization

It might be expected that the most readily available electrophilic initiator would be a proton. However, the reaction of the monomer ethene with hydrogen chloride, certainly in stoichiometric amounts, is not polymerization but straightforward electrophilic addition in which the proton first adds to the electron-rich alkene double bond, giving a carbocation, which then adds to the nucleophilic Cl^- (Fig. 3.17a).

For polymerization, the amount of H^+ must be kept low, just enough to initiate the chain mechanism. Also the anion should be non-nucleophilic to

(a) $$CH_2=CH_2 \xrightarrow{HCl} CH_3{-}CH_2^{\oplus} \xrightarrow{Cl^{\ominus}} CH_3CH_2Cl$$

Ethene ... Chloroethane

(b) $$CH_2=CH_2 \xrightarrow{HCl} CH_3{-}CH_2^{\oplus} \xrightarrow{CH_2=CH_2} CH_3CH_2CH_2CH_2^{\oplus}$$

$$\xrightarrow{\text{repeat}} \longrightarrow CH_3CH_2(CH_2CH_2)_n\,CH_2CH_2^{\oplus} \xrightarrow{Cl^{\ominus}}$$

$$\longrightarrow CH_3CH_2(CH_2CH_2)_n\,CH_2CH_2Cl$$

Chlorine terminated polymer

Fig. 3.17 Reactions of ethene with hydrogen chloride (a) stoichiometric small molecule reaction gives chloroethane; (b) low concentration of HCl allows polymerization of intermediate carbocationic species before termination by nucleophilic anion.

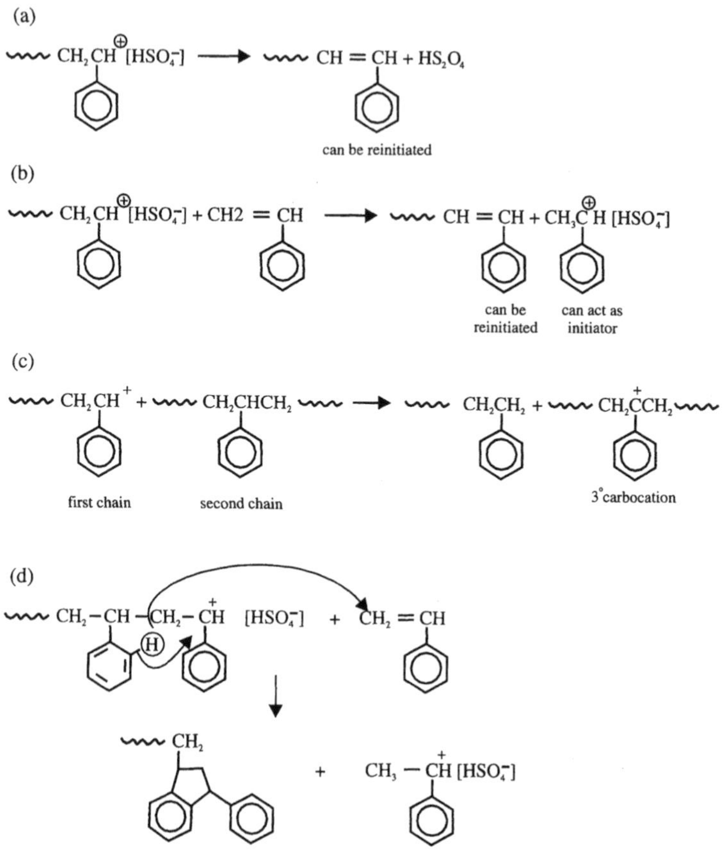

Fig. 3.18 Termination mechanisms for acid-catalysed carbocationic chain polymerization (exemplified by styrene) (a) transfer to counter-ion; (b) transfer to monomer; (c) abstraction; (d) ring alkylation. (NB: nucleophilic attack of anion not shown)

allow a decent extent of propagation before termination (Fig. 3.17b). Less nucleophilic anions than Cl^- must be used, e.g. ClO_4^-, HSO_4^- and CF_3COO^-. In this type of polymerization, one end of the polymer will be CH_3 while the other end will depend upon the termination mechanism. For example, attack of CF_3COO^- can give ester formation, but perchlorate and bisulphate esters are less stable and other termination reactions occur. If present, water will react to give hydroxy end-groups. Alternatively, loss of H^+ gives an alkene-ended chain.

Technically if H^+ is not lost and simply rejoins the anion to reform the acid catalyst (i.e. H^+A^-) then termination is termed 'transfer to counter-ion' (Fig. 3.18a), while if the H^+ initiates another monomer to start a new chain then it is termed 'transfer to monomer' (Fig. 3.18b). There could also be abstraction from another chain (Fig. 3.18c) or more complex reactions such as ring alkylation (Fig. 3.18d). This latter reaction is specific to styrene and similar ring-substituted monomers, but exemplifies the range of chemical opportunities in even a simple alkene monomer. A further complication is that aliphatic carbocations rearrange in the order primary → secondary → tertiary, which can lead to branching in the polymer.

Another means of initiating cationic polymerization, particularly in non-aqueous media, is the use of carbenium ion salts, e.g. tropylium (cycloheptatrienyl cation) or trityl chloride (Fig. 3.19). Here the three benzene rings stabilize the positive carbon atom which attacks the first monomer, giving in principle a trityl end-group. However, only when R is aromatic, OMe (ether) or a similar electron-donating species, is the alkene (CH$_2$=CHR) sufficiently reactive for this type of initiation.

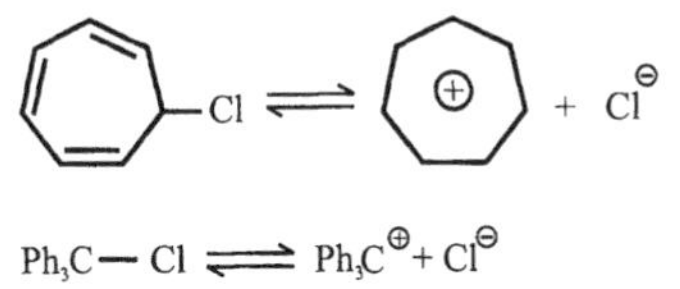

A more widely applicable non-aqueous initiation system involves Lewis acids (e.g. AlCl$_3$, BF$_3$, TiCl$_4$, FeCl$_3$). In principle, the electron-rich alkene double bond co-ordinates to the electron-deficient metal atom. However, the Lewis acid by itself is insufficient to cause polymerization and therefore small amounts of 'co-initiator' or 'co-catalyst' are added. These have an active hydrogen atom (e.g. water, alcohols) and form small amounts of a complex protic acid, e.g. H$^+$BF$_3$OH$^-$ which inserts hydrogen on the initial polymer end-group.

For example, isobutylene can also be polymerized to a very high RMM of several million in only a few seconds at $-70\,°C$ by use of a BF$_3$ catalyst and a trace of a proton donor (Fig. 3.20). Here the presence of the two electron-donating methyl groups favour cationic polymerization. A practical problem is that the reaction is exothermic and gives out enough heat to degrade the polymer. Thus it is necessary to have a heat management system which avoids temperature build-up.

General kinetics of Lewis acid catalysed cationic polymerization
In common with free radicals, the initiation step of Lewis acid catalysis is thought to take place with a slow and a fast step, i.e.

AX + BH → [H$^+$] + [BAX]$^-$ (slow step, generation of active species)
[H$^+$] + M → [MH$^+$]   (fast step, attack of first monomer).

Here AX is the Lewis acid, BH is the Lewis base and M represents monomer.

The propagation is simply the sequential addition of many monomers to the active complex [MH$^+$]:

[MH]$^+$ + M → [M$_2$H]$^+$ (i.e. [M$_2$]) or in general [M$_n$]$^+$ + M → [M$_{n+1}$]

The kinetics of the propagation process depends upon both the strength of the acid (stronger acids give more ions and faster rates) and the relative permittivity (dielectric constant) of the solvent (the larger the dielectric constant of the polymerization medium, the larger the separation of the ions and the faster the rate). For example, in order of electrophilic strength, HClO$_4$ (HClO$_4$ ⇌ H$^+$ + ClO$_4^-$) is stronger than TiCl$_4$ (TiCl$_4$ + H$_2$O ⇌ H$^+$ [TiCl$_4^-$]) which is stronger than I$_2$ (I$_2$ ⇌ I + I$_3^-$). Table 3.3 gives various propagation rate constants using the above initiator systems for the polymerization of styrene. Again it is important to note that during propagation there is no change in the number of ions. Only the initiation and termination steps alter these numbers.

Ph$_3$C — Cl ⇌ Ph$_3$C$^\oplus$ + Cl$^\ominus$

Fig. 3.19 Representative carbenium ion salts for carbocationic chain polymer initiation.

Table 3.3 Effect of dielectric constant (ε) of the polymerization medium on the styrene propagation rate constant (kp) for cationic initiation

Solvent	Relative permittivity (ε)	Catalyst system	kp/arbitrary units
CH_2Cl_2	9.7	$HClO_4$	17.0
CH_2Cl_2	9.7	$TiCl_4/H_2O$	6.0
CH_2Cl_2	9.7	I_2	0.003
CH_2Cl_2	9.7	$HClO_4$	17.0
80/20 CH_2Cl_2/CCl_4	7.0	$HClO_4$	3.20
60/40 CH_2Cl_2/CCl_4	5.2	$HClO_4$	0.40
CCl_4	2.3	$HClO_4$	0.0012

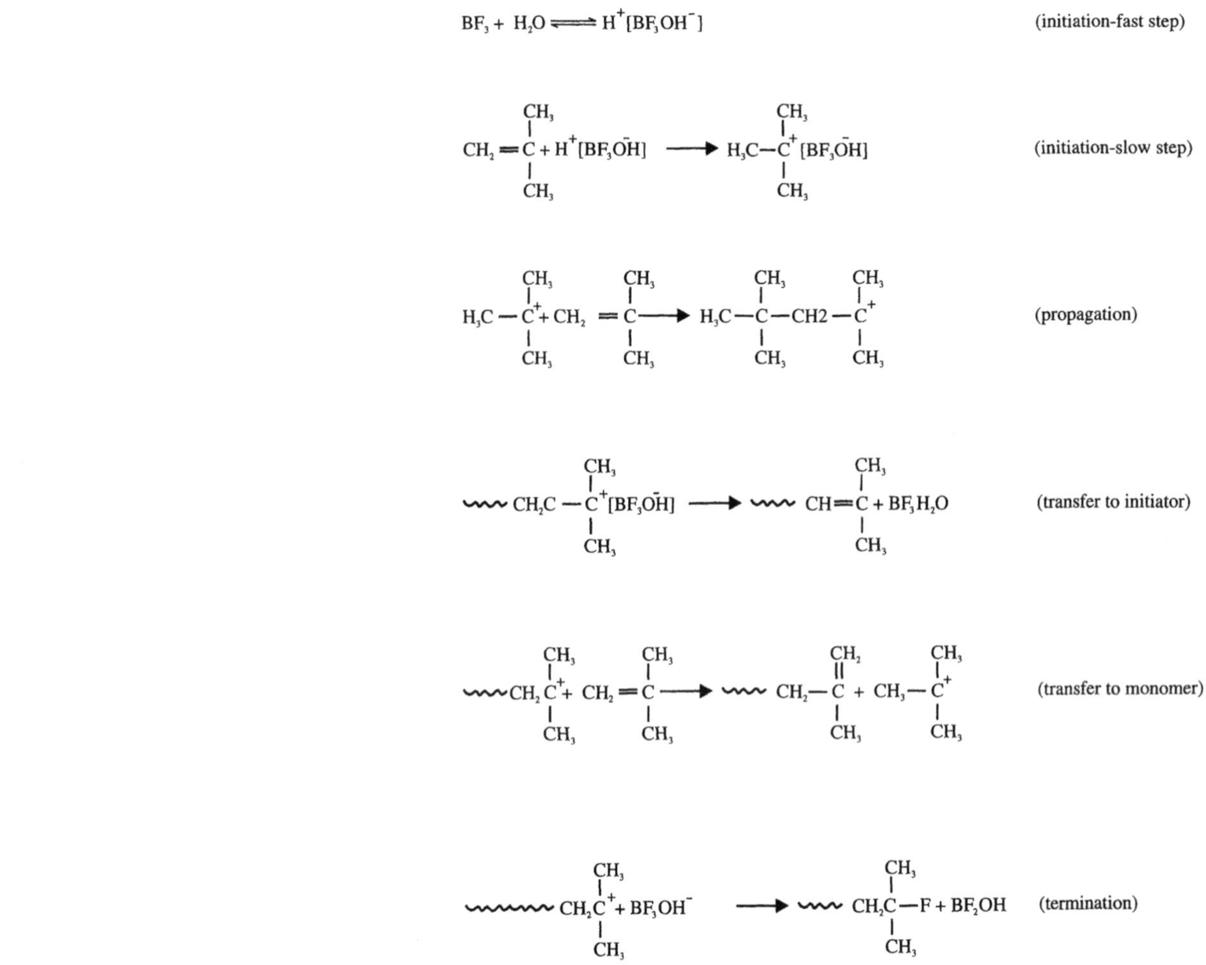

$$BF_3 + H_2O \rightleftharpoons H^+[BF_3OH^-] \qquad \text{(initiation-fast step)}$$

(initiation-slow step)

(propagation)

(transfer to initiator)

(transfer to monomer)

(termination)

Fig. 3.20 Processes occurring during cationic polymerization of isobutylene (2-methyl propene).

Kinetics of isobutene initiated by BF$_3$/H$_2$O
Initiation is given by eqn 3.25.

$$R_i = k_i[M][C] \qquad (3.25)$$

where C = catalyst/co-catalyst (BF$_3$/H$_2$O)
Propagation is given by eqn 3.26.

$$R_p = [MH^+][M] \qquad (3.26)$$

Termination by **rearrangement** is given by eqn 3.27.

$$R_t = k_t[MH^+] \qquad (3.27)$$

Applying the stationary state (i.e. $R_i = R_t$) gives eqns 3.28 and 3.29.

$$R_p = (k_p k_i/k_t)[M]^2[C] \qquad (3.28)$$

$$X_n = (k_p/k_t)[M] \qquad (3.29)$$

If termination by **transfer** to monomer is assumed, i.e. $R_{tr} = k_{tr}[MH^+][M]$, then application of the stationary state gives eqns 3.30 and 3.31.

$$R_p = (k_p k_i/k_{tr})[M][C] \qquad (3.30)$$

$$X_n = (k_p/k_{tr}). \qquad (3.31)$$

Thus, whether the synthesis of polymer by this route is terminated by either rearrangement or transfer, the dependence of rate and resultant molar mass on the initial concentrations of initiator and monomer are less complex than those observed in free radical polymerization (eqns 3.20 and 3.21).

Anionic chain polymerization
This type of process is initiated by nucleophiles to provide the end of the propagating chain with an anion. Monomers susceptible to this form of polymerization should have an electron-withdrawing R group capable of stabilizing the carbanion.

$$Nu^- + CH_2 = CHR \longrightarrow Nu - CH_2 - \bar{C}HR$$

The susceptibility of the monomer to attack also depends upon the activity of the initiator. For instance, in the series of organometallic compounds benzyl lithium is less basic (more delocalized) than ethyl lithium, and fluorenyl lithium is even less basic.

If the different monomers methyl methacrylate, styrene and propene are compared with each of these initiators we see the effect of a subtle balance of monomer – initiator reactivities (Fig. 3.22; Table 3.4).

As Table 3.4 shows, fluorenyl lithium will initiate methyl methacrylate but not styrene, benzyl lithium will initiate both, and alkyl lithium will even initiate propene.

Other examples of where the strength of the base is important are the nitroalkenes (Fig. 3.23) and the polymerization of cyanopropene (Fig. 3.24).

The monomer cyanopropene polymerizes readily in contact with moisture and the cyanoacrylate adhesives are based on such similar high reactivity.

$$< CH_2 CH_2^- \ Li^+$$
ethyl   lithium

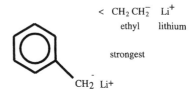

strongest

Benzyl Lithium

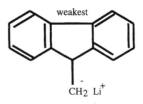

weakest

Fluorenyl lithium

Fig. 3.21 Basicity of representative anionic initiators.

Table 3.4 Comparison of different initiator/monomer systems

Increase in basicity of initiator	Increasing tendency to anionic polymerization →		
	Propene	Styrene	Methyl methacrylate
Fluorenyl lithium	✗	✗	✓
Benzyl lithium	✗	✓	✓
Ethyl lithium	✓	✓	✓

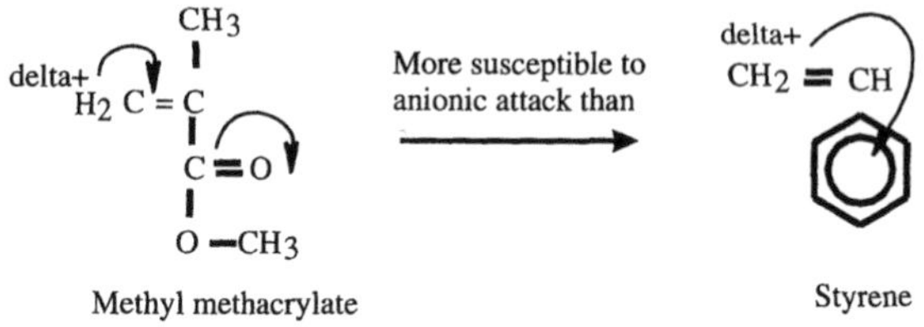

Fig. 3.22   Anionic chain polymerisation methyl methacrylate *versus* styrene.

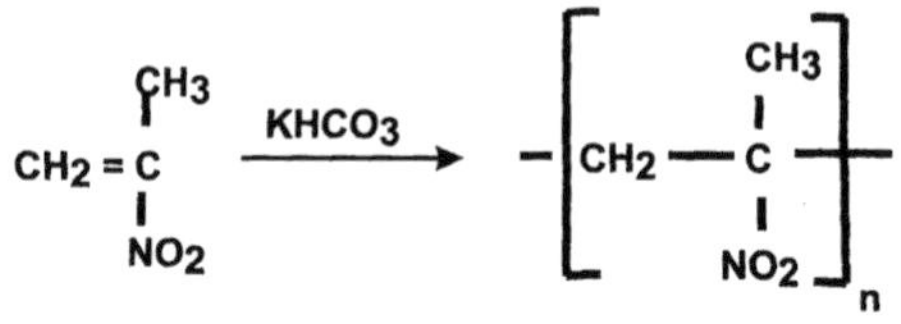

Fig. 3.23   Anionic chain polymerisation of a nitroalkene.

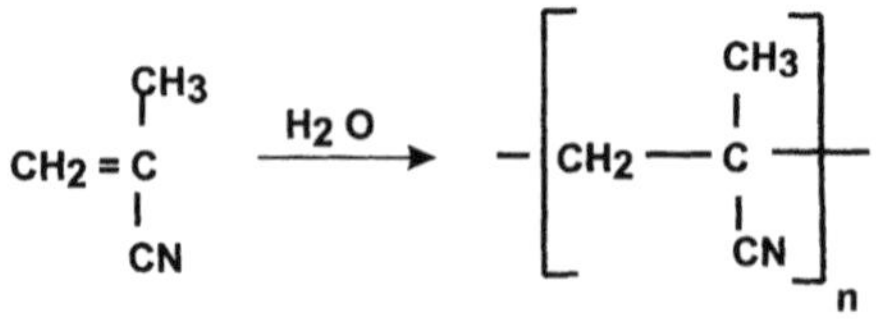

Fig. 3.24   Anionic chain polymerisation of cyanopropene.

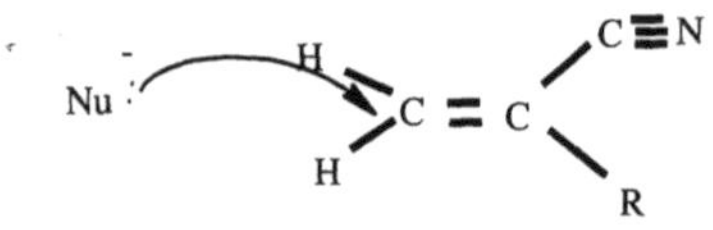

Fig. 3.25 Cyanoacrylate polymerization.

Cyanoacrylates are the principal component of 'Super Glue', which is notorious for problems associated with skin contact. The very strong electron-withdrawing effect of the ester and the cyano groups makes the monomer highly susceptible to attack by nucleophilic water present on the skin (Fig. 3.25).

The most commonly used anionic initiators are organometallic compounds, especially organolithium compounds. These are air-sensitive compounds soluble in hydrocarbon solvents and are very reactive towards compounds with labile protons, e.g. H_2O, ROH, RNH_2, RCO_2H, etc. Other powerful anionic species are salts of the type X^+Y^- where Y is NH_2 or similar. For example, potassium metal is so reactive that it will remove a proton from ammonia to evolve hydrogen and give potassium amide ($K^+NH_2^-$, 'potassamide'). The amide ion is the anionic initiator. Finally, alkali metals can react directly with alkenes by electron transfer (Fig. 3.26a), or else the process can be mediated by a molecule such as naphthalene (Fig. 3.26b). Examples of each of the initiator systems follow.

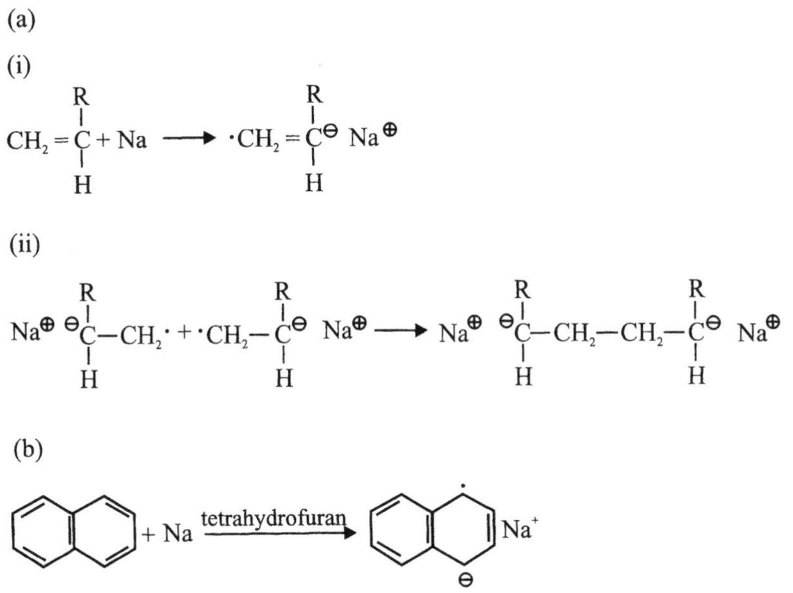

(a)

(i)

$$CH_2=\overset{\displaystyle R}{\underset{\displaystyle H}{C}}+Na \longrightarrow \cdot CH_2=\overset{\displaystyle R}{\underset{\displaystyle H}{C}}{}^{\ominus}\ Na^{\oplus}$$

(ii)

$$Na^{\oplus}\ {}^{\ominus}\overset{\displaystyle R}{\underset{\displaystyle H}{C}}-CH_2\cdot + \cdot CH_2-\overset{\displaystyle R}{\underset{\displaystyle H}{C}}{}^{\ominus}\ Na^{\oplus} \longrightarrow Na^{\oplus}\ {}^{\ominus}\overset{\displaystyle R}{\underset{\displaystyle H}{C}}-CH_2-CH_2-\overset{\displaystyle R}{\underset{\displaystyle H}{C}}{}^{\ominus}\ Na^{\oplus}$$

(b)

Fig. 3.26 Anionic initiation of chain polymerization (a i) direct initiation by metal to give radical anion; (ii) dimerization of radical anion; (b) indirect system using metal and mediator (first step shown only).

Polymerization of styrene with ethyl lithium

The production of the ethyl lithium salt is achieved by reacting lithium metal with ethyl chloride (eqn 3.32) in the presence of a non-polar solvent (e.g. hexane)

$$2Li + C_2H_5Cl \rightarrow LiC_2H_5 + LiCl \tag{3.32}$$

The actual initiation step involves the reaction of the ethyl lithium with the monomer, again in the inert solvent at room temperature (Fig. 3.27).

NB-since the styryl anion is highly coloured, the kinetics of the initiation process can be followed by spectroscopy ($\lambda = 350$ nm)]

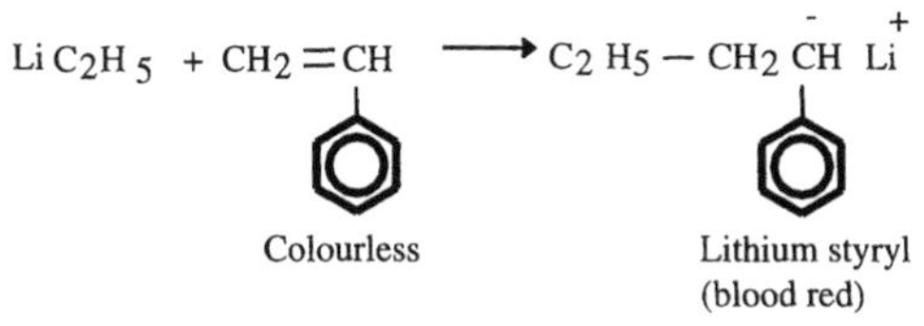

Fig. 3.27 Ethyl lithium initiating styrene polymerization.

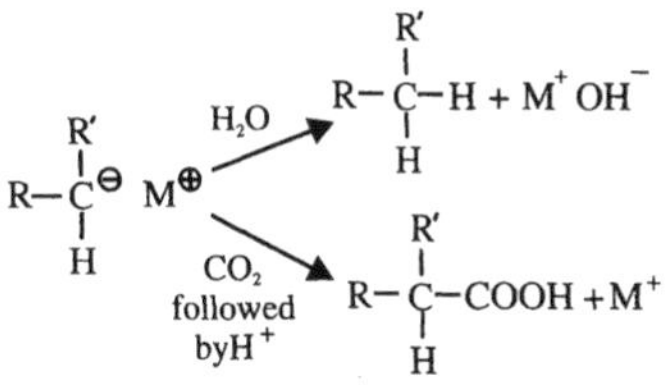

Fig. 3.28 Representative termination processes for anionically initiated chain polymerizations.

It is important that a non-polar solvent is used throughout since a lithium alkyl is very reactive. For example, the presence of water or carbon dioxide would prohibit further polymerization as shown in Fig. 3.28.

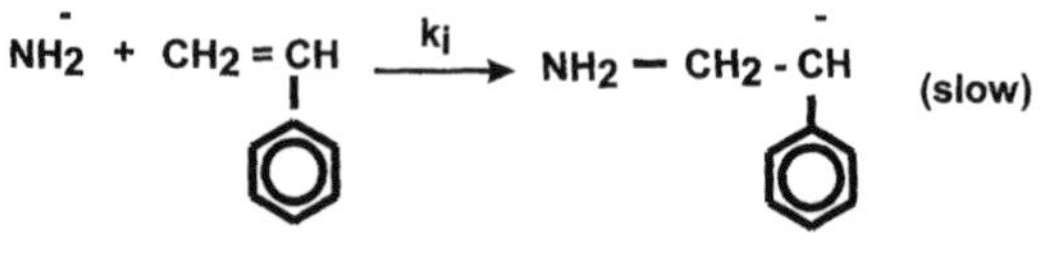

Fig. 3.29 A metal amide initiating styrene polymerization (M^+ ion not shown).

Styrene and potassium amide (KNH₂)

This reaction was the first anionic polymerization studied. Preparation of the initiator takes place by dissolving potassium in liquid ammonia ($-33\,^\circ$C). The initiation steps are given by eqn 3.33 and Fig. 3.29.

$$KNH_2 \rightleftharpoons K^+ + NH_2^- \; \text{(fast)} \tag{3.33}$$

The initiation reaction is followed by propagation (Fig. 3.30) and termination, in which the cessation of polymerisation occurs by a transfer process in which a proton is abstracted from the solvent (liquid ammonia) by the growing anion (Fig. 3.31).

Kinetically, the above three steps of initiation, propagation and termination (strictly speaking transfer since another amide ion is generated) can be represented respectively as eqns 3.34 – 3.36.

$$R_i = k_i[NH_2^-][M] \tag{3.34}$$

$$R_p = k_p[M][M^-] \tag{3.35}$$

$$R_{tr} = k_{tr}[M^-][NH_3] \tag{3.36}$$

Using the stationary state hypothesis (see Section on free radicals) allows derivation of eqns 3.37 and 3.38.

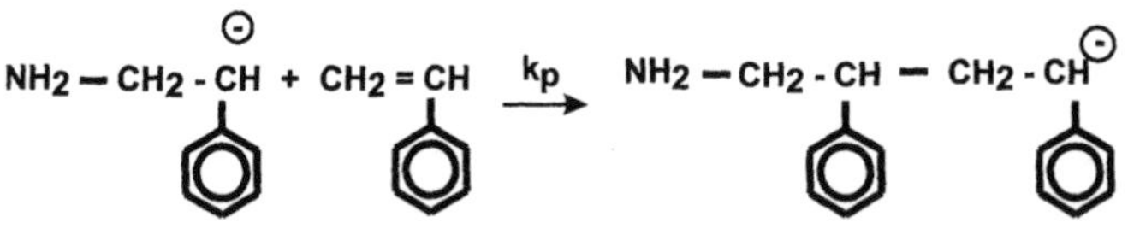

Fig. 3.30 Propagation in the amide styrene system.

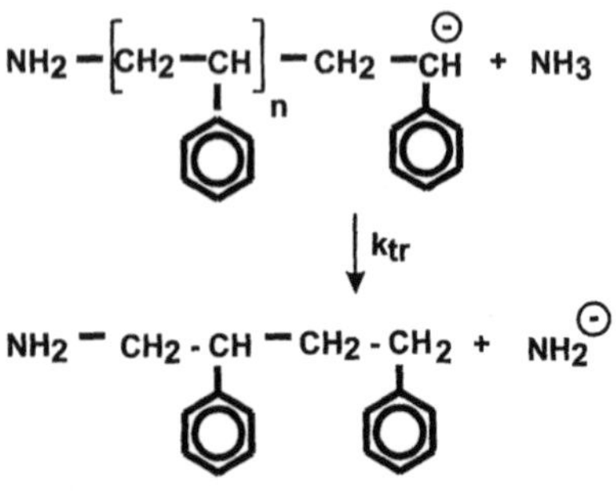

Fig. 3.31 Termination in the amide styrene system.

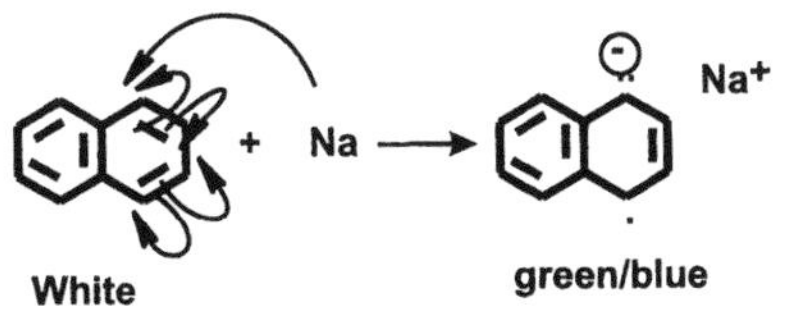
+ Na
Na+
White
green/blue

Fig. 3.32 Sodium naphthalene initiator.

$$R_p = \frac{k_p k_i}{k_{tr}} \frac{[NH_2^-][M]^2}{NH_3} \tag{3.37}$$

$$\overline{X_n} = \frac{k_p[M]}{k_{tr}[NH_3]} \tag{3.38}$$

In other words, doubling the styrene concentration leads to a doubling of the molar mass (eqn 3.38) and a fourfold increase in the polymerization rate (eqn 3.37).

Experimentally the rate of polymerization is found to conform to equation 3.37 showing the usefulness of the stationary state hypothesis in this system.

Styrene and sodium naphthalide
This particular reaction is an example of the use of naphthalene as the electron transfer agent. Initiation involves the reaction of sodium and naphthalene to produce sodium naphthalide (Fig. 3.32), which subsequently reacts with styrene to produce a dianion (Fig. 3.33), both ends of which grow by reaction with monomer.

There is an interesting situation regarding the termination of all anionic chain polymerizations. Carbanions have completely filled orbitals and do not rearrange. Carbocations on the other hand have empty orbitals into which electron density can flow as bonds move, which is why they rearrange. Also, loss of a hydride ion (H^-) as a high energy species is disfavoured by thermodynamics. Thus termination is mostly via transfer to solvent. However,

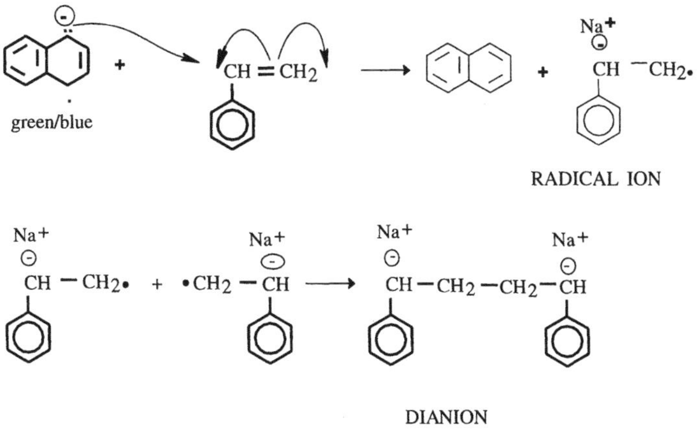
green/blue
CH = CH2
Na+
CH —CH2•
RADICAL ION
Na+
CH — CH2•
+
•CH2 — CH
CH—CH2—CH2—CH
DIANION

Fig. 3.33 Production of dianion.

in many cases rigorous exclusion of water, oxygen, carbon dioxide or reactive impurities leads to polymerization with *no* inherent termination or chain transfer. Thus, when styrene is polymerized by sodium naphthalide, the green–blue colour of the initiator changes to red (the colour of the styrene anion) and the red colour persists even after 100% conversion. If further monomer is added the polymerization continues. This type of system is called a **living polymer**. If a second monomer is added then a **block copolymer** is obtained. When the second monomer is used up, a third monomer is added and so on.

If the living polymer is quenched with carbon dioxide, this results in a polymer with COOH end-groups (Fig. 3.28). Such a system can be included in formulations for step-growth polymerization and hence the incorporation of blocks of chain polymerization units into a step-growth system (telechelic polymers).

An important consequence of having no termination or chain transfer is that the molar mass distribution is very narrow (monodisperse), allowing the resultant polymers to be used as standards for gel permeation chromatography in the measurement of molar mass.

Other living chain polymer systems

The ready avoidance of quenching electrophiles makes anionic-initiated 'living' polymers simple to study; but other 'living' chain polymer systems are feasible with care. Thus in the absence of nucleophiles and with a chain end that does not readily rearrange or lose a proton, 'living cationic' polymers are possible, e.g. α methyl styrene or vinyl ethers. If the unpaired electron on a radical chain end is protected against combination termination by equilibrium with a spin trap then 'living radical' polymerization occurs, giving a monodisperse product with fine control of RMM. This is useful for example for methacrylates.

3.3.4 Copolymerization

This is the production of polymers from more than one monomer. They can be prepared by reacting monomer A and monomer B to give:

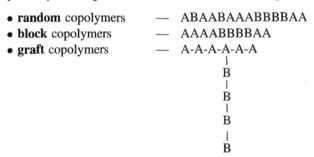

- **random** copolymers   — ABAABAAABBBBAA
- **block** copolymers   — AAAABBBBAA
- **graft** copolymers   — A-A-A-A-A-A
                                             |
                                             B
                                             |
                                             B
                                             |
                                             B
                                             |
                                             B
- **alternating** copolymers   — ABABABAB.

Copolymers are normally produced to give improved properties. For example, copolymerization of styrene with methyl methacrylate improves heat resistance (polystyrene softens at $+100\,°C$, perspex at $+120\,°C$); copolymerization of amorphous vinyl acetate with crystalline polyethene improves the flexibility; copolymerization of vinyl acetate with vinyl chloride improves the processability

Table 3.5 Approximate reactivity ratios for important chain-growth copolymers

Monomer 1	Monomer 2	Temperature (°C)	r_1	r_2
Methyl methacrylate	Acrylonitrile	80	1.20	0.15
Methyl methacrylate	Butadiene	90	0.25	0.75
Styrene	Butadiene	60	0.80	1.40
Styrene	Methyl methacrylate	80	0.50	0.50
Styrene	Vinyl acetate	60	55	0.01
Vinyl acetate	Acrylonitrile	70	0.07	6.0
Vinyl acetate	Methyl methacrylate	60	0.015	20

of PVC, which otherwise tends to degrade on heating. Often, copolymers exhibit a combination of the best properties of their components, while a simple physical blend of the polymers may emphasize less favourable aspects.

Random copolymers
These are usually prepared by free radical polymerization, e.g. Buna-S from butadiene and styrene. Styrene gives the copolymer hardness and mechanical stability while butadiene gives it flexibility and the ability to cross-link. Another example is that produced from acrylonitrile (Courtelle fibre) and vinyl pyridine. The vinyl pyridine is included to improve the dyeability of the material since commercial dyes interact with the lone pair on the pyridine nitrogen atom.

The properties of a copolymer can be varied by altering the ratio of component monomers. For example, 100% polyvinyl acetate (PVA) with a molar mass of approximately 10 000 is used chiefly for adhesives, 100% polyvinyl chloride (PVC) is used as a rubber substitute after the addition of plasticizer, while a 90:10 PVC:PVA copolymer is a plastic with good strength and solvent resistance.

A feature of chain copolymers is that the proportion of monomers in the initial starting mixture may not be reflected in the composition of the resulting polymer. In fact this will only the case if both monomers are of equal reactivity. However, assumption of the steady state hypothesis and that only propagation steps consume monomer, leads to eqn 3.39.

$$\frac{F_1}{F_2} = \frac{f_1(r_1 f_1 + f_2)}{f_2(r_2 f_2 + f_1)} \tag{3.39}$$

where f_1 and F_1 are the mole fraction of monomer 1 in the feed and the copolymer respectively. Table 3.5 gives some values of reactivity ratios for important commercial systems.

Block copolymers
These are prepared by linking together two or more long linear sequences of different homopolymers. They are usually either prepared from living chain polymers or from step-growth systems. Typical examples include 'diblock', 'triblock' and 'sandwich'. An example of a step-growth diblock is that prepared from polyurethane and polyether. The polyurethane part provides the hard crystalline segment of high melting point (T_m) and the polyether part the soft rubber segments. The combination is used to combine hardness (i.e. support) and elasticity (i.e. give) and finds use typically in 'Spandex' swimsuits which require improved stretch and shape recovery.

An example of a Sandwich copolymer is that between butadiene and styrene. The properties of the resultant materials are dependent on the filling. For example, if the filling is styrene then the material is a rigid elastomer (i.e. more strength) while if the filling is butadiene, the material is a flexible plastic, i.e. becomes less brittle and less likely to crack.

Triblock copolymers such as ABS (acrylonitrile, butadiene, styrene) have outstanding impact and mechanical strength (they are used in protective helmets) and possess good chemical resistance.

Graft copolymer

These are branched block copolymers and hence the properties are similar. A typical example is the grafting of acrylate onto a nylon backbone to improve the dyeability of the textile. Preparation includes forming a polymer backbone with pendant groups which are active to the second monomer or treating a homopolymer with ionizing radiation (electron beam) in the presence of the second monomer.

Alternating copolymers

These are completely regular (–ABABABAB–) and special conditions are necessary for their production. They are virtually impossible to produce by a free radical route and at present are of limited commercial value.

3.4   Modern chain polymer developments

Co-ordination complex catalysts

Before 1953 ethene was polymerized by a high pressure (1000 atmospheres) and moderate temperature (up to 300 °C) free radical process. Backbiting (Fig. 3.10 caused the product polymer to have a highly branched backbone and thus to have only moderate thermal and mechanical properties with limited crystallinity known as LDPE (low density polyethene).

In 1953 Ziegler found that polymerization took place at room temperature and atmospheric pressure in the presence of a variable valency transition metal catalyst. The product was a crystalline solid of empirical formula CH_2, which was free from branching and defects, and which is now called high density polyethene (HDPE).

When in 1954 Natta applied similar catalysts to the polymerization of propene, a stereoregular isostatic polymer was obtained which was much tougher than the free radical product and could be used as plastic sheeting, replacements for rope, e.g. baler twine used in farming and, when defect-free, as car panels and in other engineering applications.

Typical Ziegler–Natta catalysts have many variants and are prepared by reacting an alkyl of a metal from groups I–III (e.g. triethylaluminium, diethylzinc, etc.) with compounds of a transition metal from groups IV–VIII (e.g. titanium tetrachloride, vanadium oxychloride, molybdenum pentachloride, tungsten hexachloride and similar). The two components are dissolved in a hydrocarbon solvent (toluene or n-heptane) at room temperature to give an exothermic reaction in which gases are evolved and a dark-coloured solid, the catalyst, generally precipitates.

The reactions need care in the laboratory because of the reactivity of the reagents. For example, for a typical catalyst system triethyl aluminium is pyrophoric (spontaneously inflammable with aerial oxygen), titanium tetrachloride spontaneously hydrolyses upon exposure to water vapour (in the air) and the alkene monomers are flammable. However, industrial reactors can tolerate these materials. Fig. 3.34 indicates catalyst-forming reactions. The presence of butane, ethane, ethene, and short-chain polyethene in the catalyst mixture before the addition of the alkene monomer is due to the production of ethyl radicals.

The important chemical step appears to be the reduction of titanium (IV) to lower valencies. These are key reactive species because they can co-ordinate alkene double bonds, the first step in the polymerization process when the monomer is exposed to the pre-prepared and aged catalyst. In general it is thought that polymerization is a surface process which accounts for the stereochemical control.

There are a number of proposed mechanisms. An early one which shows the general principles was Natta's bimetallic mechanism (Fig. 3.35).

$$(C_2H_5)_3Al + TiCl_4 \longrightarrow (C_2H_5)_2AlCl + C_2H_5TiCl_3$$

$$C_2H_5TiCl_3 \longrightarrow \cdot C_2H_5 + TiCl_3$$

$$(C_2H_5)_3Al + TiCl_3 \longrightarrow (C_2H_5)_2AlCl + C_2H_5TiCl_2$$

$$C_2H_5TiCl_2 \longrightarrow \cdot C_2H_5 + TiCl_2$$

$$(C_2H_5)_3Al + C_2H_5TiCl_3 \longrightarrow (C_2H_5)_2AlCl + (C_2H_5)_2TiCl_2$$

$$(C_2H_5)_2TiCl_2 \longrightarrow \cdot C_2H_5 + C_2H_5TiCl_2$$

Fig. 3.34 Possible reactions during preparation of a co-ordination complex (Ziegler–Natta) catalyst from triethylaluminium and titanium tetrachloride.

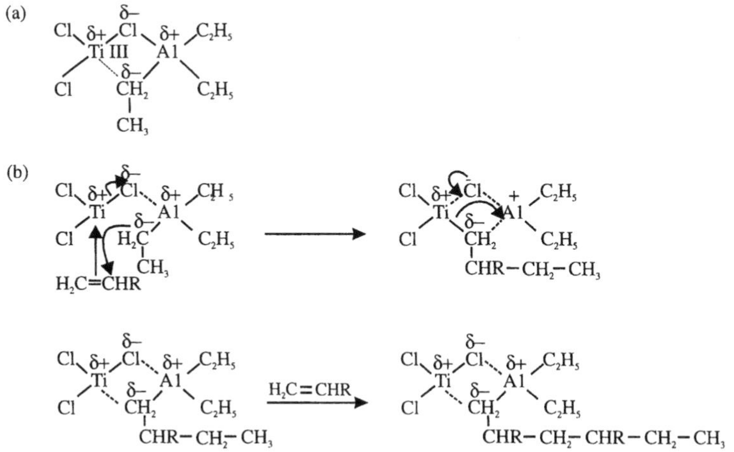

Fig. 3.35 Natta's bimetallic mechanism for co-ordination complex catalysis of chain polymerization (a) proposed catalyst complex after initial exchange reactions before monomer ingress; (b) proposed mechanism for polymerization reactions in presence of monomer (note attack of δ anionic charge onto alkene double bond).

Table 3.6 Effect of co-ordination complex catalyst upon stereo regularity of polypropene. (Free radical polymerization does not give a stereo regular product)

Catalyst	Stereo regularity (%)
$R_3Al + TiCl_4$	35
$R_3Al + TiCl_4 + NaF$	97
$R_3Al + TiCl_3$	85
$R_2AlCl + TiCl_3$	99
$RNa + TiCl_4$	90
$RLi + TiCl_4$	90

Table 3.7 Effect of initiation system upon structure of polybutadiene

Initiation	cis-1,4-	trans-1,4-	1,2-
Free radical	19	60	21
Na metal	10	25	65
Li metal	35	52	13
Butyl lithium	33	55	12
TiI_4/Al t-Bu_3	95	2	3
$CoCl_2/AlEt_2Cl$/pyridine	98	1	1
VCl_3-$AlEt_3$	0	99	1

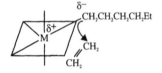

Fig. 3.36 Cossee's monometallic mechanism for co-ordination complex catalysis. New alkene approaches vacancy in metal co-ordination and is attacked by δ^-.

However, this has been replaced by other mechanisms such as Cossee's monometallic mechanism; Fig. 3.36 gives a schematic form. In Fig. 3.36, the Ti (III) species has a vacancy in its co-ordination sphere into which the alkene first co-ordinates through the π bond before reacting with the anionic end of the alkyl unit to give a σ bond to the Ti. The vacancy shifts and as the next monomer approaches for co-ordination, followed by addition to the chain, the vacancy flips again. Thus the vacancy shuttles back and forth as monomers are added to the metal atom end of the growing chain. The presence of R groups in the monomer and the symmetry and geometry requirements of the appropriate metal orbitals cause the monomer to add and 'open up' in the same way, giving the stereochemical control necessary to produce isotactic polymers. The central metal atom needs to form both σ and π bonds to carbon and this explains why iron is not a good catalyst since it prefers to form π bonds rather than σ bonds.

For syndiotacticity, the monomers must add to open up so that the substituent group alternates along the backbone chain. This is a less common situation and requires soluble catalysts to allow monomer approach from alternate directions. Vanadyl species with organic ligands such as acetoacetonate are generally used.

Table 3.6 and 3.7 show how the choice of catalyst can influence the products. Note for a diene monomer, there are several isomeric products (Fig. 3.37) and these can be distinguished by choice of Ziegler–Natta catalyst.

Metathesis polymerization

A further variant on chain polymerization concerns metathesis, which is a bond reorganization reaction related to pericyclic (frontier orbital) reactions such as cycloadditions (e.g. the Diels–Alder reaction and electrocyclic rearrangements) which involve the interconversion of π and σ bonds by orbital overlap.

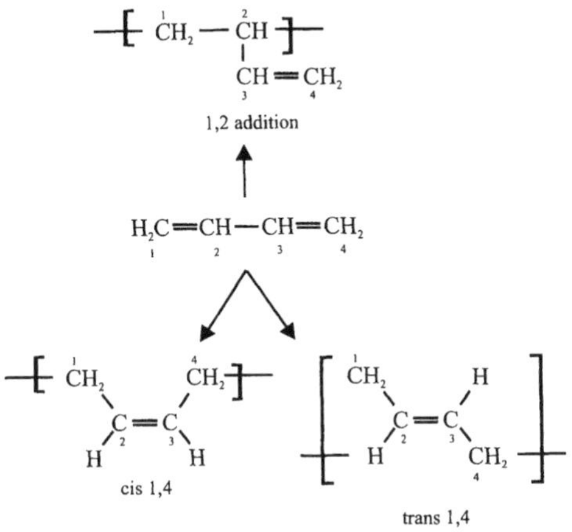

Fig. 3.37 Polymerization of butadiene to isomeric products.

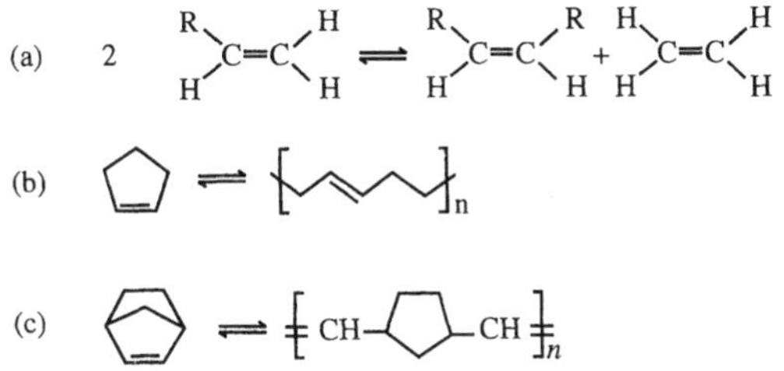

Fig. 3.38 Examples of metathesis reactions (a) generic exchange reaction in small molecule alkene; (b) formation of polypentenamer; (c) norbornene polymerization.

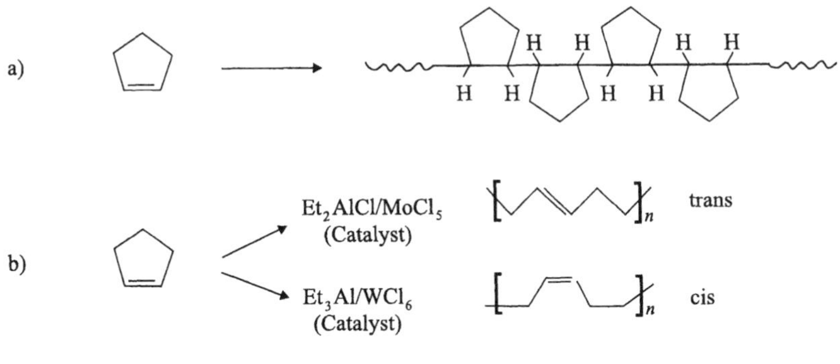

Fig. 3.39 Polymerization of cyclopentene (a) expected product from traditional alkene polymerization; (b) ring opening metathesis with control of stereochemistry.

Double bond migration in metathesis is useful in the polymerization of cyclic alkenes where opening of the ring reduces strain (Fig. 3.38). The expected product from a traditional chain polymerization of cyclopentene is sterically hindered, in contrast with the product of ring-opening metathesis polymerization (ROMP) (Fig. 3.39). The catalysts used are variations on co-ordination complexes. Minor changes in catalyst make-up can profoundly alter the product stereochemistry.

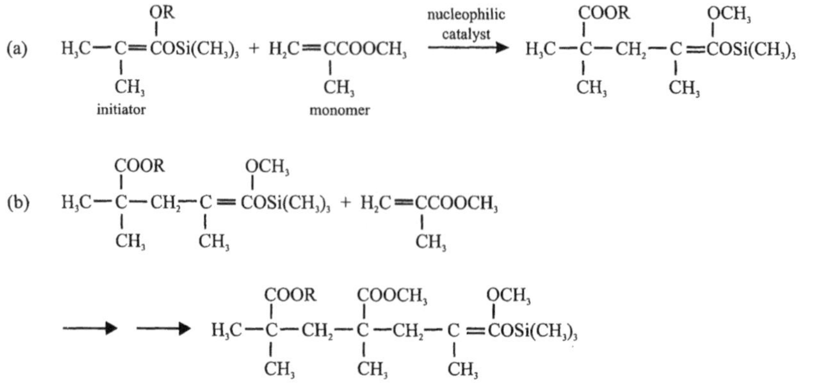

(a)   H₃C—C=COSi(CH₃)₃ + H₂C=CCOOCH₃ →[nucleophilic catalyst]→ H₃C—C—CH₂—C=COSi(CH₃)₃

with OR, CH₃, initiator on the first; CH₃, monomer on the second; COOR, CH₃, OCH₃, CH₃ on the product.

(b)   H₃C—C—CH₂—C=COSi(CH₃)₃ + H₂C=CCOOCH₃

with COOR, CH₃, CH₃, OCH₃ substituents, and CH₃ on the monomer.

→ → H₃C—C—CH₂—C—CH₂—C=COSi(CH₃)₃

with COOR, CH₃; COOCH₃, CH₃; OCH₃, CH₃ substituents.

Fig. 3.40   Group transfer polymerization (a) initiation; (b) propagation.

Group transfer polymerization

This recent development, which is a type of living polymer system, is currently somewhat expensive because rigorous purity of materials and anhydrous conditions are essential.

The monomer requires a side-group carbonyl or nitrile functionality, so the most widely used monomers are acrylates, and essentially the group transferred is a silyl ketene acetal, using a soluble nucleophilic catalyst such as cyanide, azide or fluoride salt such as bis (dimethyl amino) sulphonium bifluoride at room temperature. Initiation and propagation are shown in Fig. 3.40. Termination is by addition of an active hydrogen compound. The possibility of use for controlled coating of optical fibres with materials of well defined refractive index offers a particular benefit of this type of polymerization.

3.5   Commercial chain polymer syntheses

These include the largest tonnage thermoplastics worldwide, polythene, polystyrene, polyvinyl chloride (PVC), polyvinyl acetate and derivatives, a range of polyacylates and derivatives, artificial rubbers (polydienes), as well as speciality polymers such as polytetrafluoroethene (PTFE). There are four main methods of producing polymers, namely bulk, solution, suspension and emulsion. There are also some interesting economic and chemical aspects in the production of monomers.

Bulk polymerization

This is the simplest of the techniques and consists of mixing a liquid monomer, a chain-transfer agent (to control RMM) and a monomer-soluble initiator together. Although this gives the purest polymer, the last traces of monomer are difficult to remove. Also, the free radical polymerization is so exothermic that heat build-up becomes a problem, especially when viscosity becomes high. Careful engineering and the use of multistage processes are required. As polymerization proceeds, the diffusion rate of the polymer chains becomes less than that of the monomer and hence the rate of termination decreases, producing an autoacceleration phenomenon known as the Trommsdorff–Norrish gel effect.

Bulk polymerization is now only used for polystyrene, polymethyl methacrylate and high-pressure free radical polythene, particularly for casting formulations and low RMM adhesives, plasticizers and lubricants.

Solution polymerization

Both the initiator and the monomer must be soluble. While the heat sink offered by the solvent makes temperature control easier during reaction, factors such as solvent toxicity, flammability, initial cost, boiling point and reactivity can restrict the choice of system, and chain transfer to solvent can put a limit on the polymer RMM attainable. It can also be difficult to remove the last traces of solvent from the product.

Polymers prepared in this way from aqueous media include polyvinyl pyrrolidone, polyacrylamide and polyacrylic acid, while polymers prepared from organic solvents (e.g. hydrocarbons, ethers, esters and alcohols) include polystyrene, polyvinyl chloride, polybutadiene, polymethyl methacrylate and polyvinylidene fluoride. The correct choice of system produces a solution of polymer of sufficient concentration to be processed directly or used directly onwards for certain adhesives and paints.

Suspension polymerization

Immiscible liquids form suspensions of microdroplets of diameter $\sim 10^{-2}$ cm under sufficient mechanical agitation, with each droplet essentially being a small scale bulk polymerization reactor. Although water is a good medium for heat dissipation and thermal management, the initiator must be soluble in the monomer. An advantage of this process is that the polymer is often obtained as small beads which are easily collected. Solvent removal is simple, and while the recovered polymers are relatively pure, polymers which are soluble in their own monomers become tacky and hence gelatine, polyvinyl alcohol, methyl cellulose and other stabilizers should be added.

Typical polymers prepared in this way include polystyrene, polyvinyl chloride and PTFE. PTFE is used for its very low friction coefficient as non-stick coatings and also because of its great thermal stability and resistance to chemical attack. However, it is a very insoluble polymer and must be powder-processed, sintered or processed as a paste.

Emulsion polymerization

This is the most widely used commercial process for free radical vinyl and diene polymerizations. The system consists of water as the heat transfer medium, a water-soluble initiator (e.g. potassium or ammonium persulphate), an immiscible monomer (capable upon agitation of forming droplets of diameter $\sim 10^{-2}$ cm) and a surfactant (i.e. a soap-like species) such as sodium dodecylbenzene sulphonate (e.g. $C_{12}H_{23}PhSO_3^-$) to stabilize the droplets in solution.

A latex is formed consisting of polymer particles between 0.05 and 2 μm in diameter. It is stable and may be used directly as latex paints or may be

Table 3.8 Typical rubber/latex recipes and production time

Ingredients/ mass	Styrene–butadiene rubber	Polyacrylate latex
Water	190	133
Butadiene	70	–
Styrene	30	–
Ethylacrylate	–	93
2-chlorovinyl ether	–	5
p-divinyl benzene	–	2
Soap	–	3
$K_2S_2O_8$ (initiator)	0.3	1
n-Dodecylmercaptan	0.5	–
Sodium pyrophosphate	–	7
Reaction time (h)	12	8
Temperature (°C)	50	60
Yield (%)	65	100

Table 3.9 Comparison of techniques for free radical polymerization

Method	Advantages	Disadvantages
Bulk	Simple No added contaminants	Reaction exothermic Difficult to control High viscosity
Suspension	Heat rapidly dispersed Low viscosity Polymer obtained in granular form May be used directly	Washing or drying may be needed Agglomeration Contamination by stabilizer
Solution	Heat rapidly dispersed Low viscosity May be used directly as a solution	Cost of solvent Solvent difficult to remove Possible environmental pollution
Emulsion	Heat easily dispersed Low viscosity May be used directly as emulsion Works for tacky polymers	Contamination by emulsifier Chain transfer agents needed to control RMM Washing/drying needed

coagulated (by increasing ionic strength by the addition of salt or acid), stripped of residual monomer and spray-dried. The technique is very useful, but the polymers contain surfactants which interfere with electrical insulation or optical clarity.

Polyacrylates are also prepared by emulsion methods. This method is preferred because of the high exothermicity produced during bulk polymerization, the tendency of soft particles to coalesce during suspension polymerization and the viscosity problems which arise during solution polymerization. Emulsions are generally not affected by viscosity factors, i.e. lattices of high RMM polymers do not become overly viscous. Typical emulsion polymerization systems for a rubber or a latex are given in Table 3.8. A comparison between the various commercial free radical chain polymerization methods is given in Table 3.9.

3.6 Synthesis of monomers

Monomers based on ethene and ethyne

Styrene (phenylethene) is produced by a Friedel–Craft reaction from ethene (obtained from catalytic cracking of petroleum) and benzene, both cheap, to give initially ethylbenzene which is subsequently catalytically dehydrogenated.

Before the 1960s and the ready availability of petroleum and natural gas, *vinyl chloride* and several other vinyl monomers were prepared from acetylene (ethyne). Ethyne was obtained from chalk and limestone (calcium carbonate) by heating with carbon to give first calcium carbide and then treating with water to produce ethyne gas.

More recently, ethyne has been obtained as a product of catalytic cracking and reforming of natural gas according to Fig. 3.41b, although the procedure is more complex.

Production of vinyl chloride (or other vinyl monomers) requires the addition of hydrochloric acid (or an appropriate acid) to ethyne (Fig. 3.42a). Optimization of the process is necessary (appropriate catalysts and temperature control) to avoid a second addition to the remaining unsaturated bond (Fig. 3.42b).

Monomers that were originally prepared from ethyne (until the 1960s) also include vinyl acetate (ethenylethanoate) as well as vinyl chloride. Ethene is now relatively cheap, and these monomers are prepared as shown in Figs 3.43 and 3.44 respectively.

There is an interesting economic point in vinyl chloride preparation, since chlorine (prepared by electrolysis of NaCl) is a relatively expensive feedstock. The reaction in Fig. 3.44 (b) (i) wastes one chlorine atom as an HCl byproduct, while the reaction in Fig. 3.44 (b) (ii) employs HCl as the main reagent. So producers operate a 'balanced' route in which reactions (a) and (b) are run in parallel for economical usage of both chlorine atoms from Cl_2.

The general scheme for the formation of a substituted alkene from ethene by oxidative substitution is shown in Fig. 3.45.

(a) $CaCO_3 \xrightarrow{\Delta} CaO + CO_2 \nearrow$

$CaO + 3C \xrightarrow{3000°C} CaC_2 + CO \nearrow$

$CaC_2 + H_2O \longrightarrow HC \equiv CH + Ca(OH)_2$

(b) $2CH_4 \underset{\rightleftharpoons}{\overset{1500°C}{}} HC \equiv CH + 3H_2$

Fig. 3.41 Commercial preparation of ethyne (a) old route via calcium carbide; (b) newer route by cracking of natural gas.

a) $HC \equiv CH + HX \longrightarrow H_2C = CHX$

b) $H_2C = CHX + HX \longrightarrow H_3C - CHX_2$

Fig. 3.42 Representative formation of vinyl monomer from ethyne (a) desired reaction, single addition; (b) over-reaction, second addition.

(a) (i) $HC \equiv CH + CH_3COOH \xrightarrow[200°C]{Zn\ salt} H_2C=CH-O-\overset{\overset{\displaystyle O}{\|}}{C}-CH_3$

(ii) $H_2C=CH-OCOCH_3 + CH_3COOH \longrightarrow H_3C-CH(OCOCH_3)_2$

(b) $H_2C=CH_2 + CH_3COOH + \tfrac{1}{2}O_2 \xrightarrow[\substack{200°C \\ 10\ atmospheres}]{Pd\ catalyst} H_2C=CH-O-\overset{\overset{\displaystyle O}{\|}}{C}-CH_3 + H_2O$

Fig. 3.43 Commercial syntheses of vinyl acetate (ethenyl ethanoate) (a) old route from ethyne (i) desired reaction; (ii) over-reaction; (b) newer route from ethene.

(a) $HC≡CH + HCl \xrightarrow[\substack{100°C \\ 1.5 \text{ atmos}}]{HgCl_2} H_2C=CHCl$

(b)   (i) $H_2C=CH_2 + Cl_2 \xrightarrow[\substack{60°C \\ 4 \text{ atmos}}]{FeCl_3} ClH_2C-CH_2Cl \xrightarrow[\substack{25 \text{ atmos}}]{500°C} H_2C=CHCl + HCl$

(ii) $H_2C=CH_2 + 2HCl + \frac{1}{2}O_2 \xrightarrow[\substack{200°C \\ 10 \text{ atmos}}]{Cu^I/Cu^{II}} ClH_2C-CH_2Cl + H_2O \xrightarrow[\substack{25 \text{ atmos}}]{500°C} H_2C=CHCl + HCl$

Fig. 3.44   Synthesis of chloroethene (vinyl chloride) from (a) ethyne; (b) ethene (i) chlorination route; (ii) oxychlorination route.

$$H_2C=CH_2 + HX + [O] \longrightarrow H_2C=CHX + H_2O$$

Fig. 3.45   Representative formation of vinyl monomer from ethene by oxidative substitution.

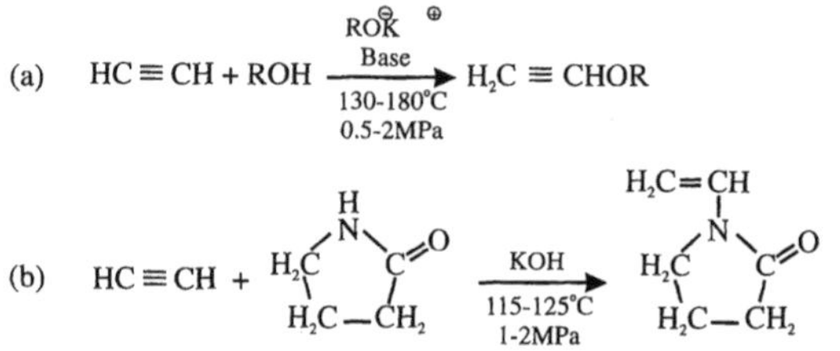

(a) $HC≡CH + ROH \xrightarrow[\substack{130-180°C \\ 0.5-2MPa}]{\substack{RO^{\ominus} \oplus \\ \text{Base}}} H_2C≡CHOR$

(b) $HC≡CH +$ [pyrrolidone structure] $\xrightarrow[\substack{115-125°C \\ 1-2MPa}]{KOH}$ [n-vinyl pyrrolidone structure]

Fig. 3.46   Syntheses from ethyne:- (a) a vinyl ether; (b) *n*-vinyl pyrrolidone.

There still remain some chain monomers which are more economically prepared from ethyne. These include *vinyl ethers* and *N-vinyl pyrrolidone*. For *vinyl ethers* (Fig. 3.46a), base catalysed addition of the appropriate alcohol across the triple bond suffices. These monomers are best polymerized by cationic initiators such as $BF_3.H_2O$. *Vinyl pyrrolidone* can also be prepared by the base catalysed addition of the lactam pyrrolidone (i.e. cyclic amide) to the triple bond of ethyne (Fig. 3.46b). The monomer is water-soluble and aqueous solution polymerization is employed with hydrogen peroxide initiator to give the water-soluble polymer.

Other vinyl synthetic routes

In principle many other vinyl monomers could be produced from ethyne or ethene, e.g. acrylic acid (propenoic acid, Fig. 3.47a and b) or acrylonitrile (cyanoethene, Fig. 3.48a and b). However, both monomers are now prepared from the relatively cheap monomer propene (Fig. 3.47c and 3.48c). Note the use of catalysts and the forcing conditions required to produce the monomers.

The monomer *methyl methacrylate* is synthesized from (cheap) acetone by first reacting with HCN and second with sulphuric acid, which dehydrates to give the double bond, while at the same time causing hydrolysis of the cyano group first to the amide (which is not isolated) and then to the acid (Fig. 3.49). The various esters can be made from the methacrylic acid by esterification.

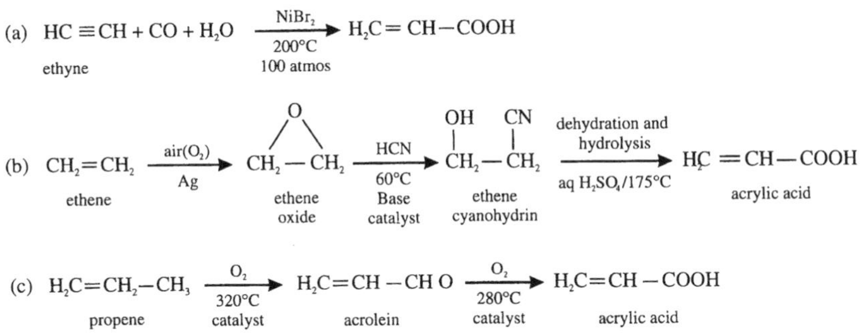

(a) $HC \equiv CH + CO + H_2O \xrightarrow[\substack{200°C \\ 100\ atmos}]{NiBr_2} H_2C = CH - COOH$

ethyne

(b) $CH_2 = CH_2 \xrightarrow[Ag]{air(O_2)} \underset{\substack{\text{ethene} \\ \text{oxide}}}{CH_2 - CH_2} \xrightarrow[\substack{60°C \\ Base \\ catalyst}]{HCN} \underset{\substack{\text{ethene} \\ \text{cyanohydrin}}}{\overset{OH\quad CN}{CH_2 - CH_2}} \xrightarrow[aq\ H_2SO_4/175°C]{\substack{\text{dehydration and} \\ \text{hydrolysis}}} \underset{\text{acrylic acid}}{HC = CH - COOH}$

ethene

(c) $\underset{\text{propene}}{H_2C = CH_2 - CH_3} \xrightarrow[\substack{320°C \\ catalyst}]{O_2} \underset{\text{acrolein}}{H_2C = CH - CHO} \xrightarrow[\substack{280°C \\ catalyst}]{O_2} \underset{\text{acrylic acid}}{H_2C = CH - COOH}$

Fig. 3.47 Synthetic routes to acrylic acid (a) route from ethyne; (b) route from ethene oxide (oxirane); (c) route from propene.

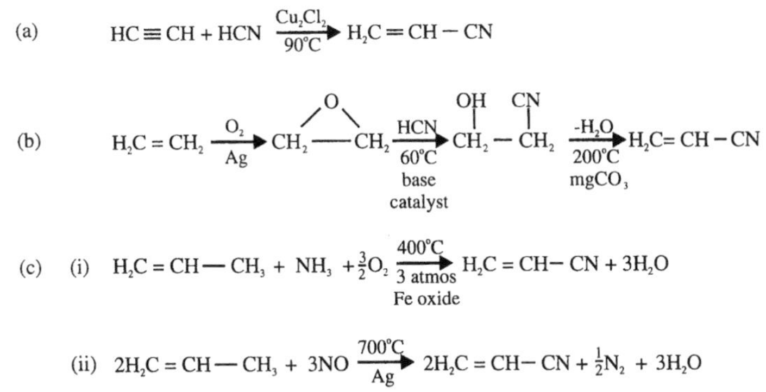

(a) $HC \equiv CH + HCN \xrightarrow[90°C]{Cu_2Cl_2} H_2C = CH - CN$

(b) $H_2C = CH_2 \xrightarrow[Ag]{O_2} CH_2 - CH_2 \xrightarrow[\substack{60°C \\ base \\ catalyst}]{HCN} \overset{OH\quad CN}{CH_2 - CH_2} \xrightarrow[\substack{200°C \\ mgCO_3}]{-H_2O} H_2C = CH - CN$

(c) (i) $H_2C = CH - CH_3 + NH_3 + \tfrac{3}{2}O_2 \xrightarrow[\substack{3\ atmos \\ Fe\ oxide}]{400°C} H_2C = CH - CN + 3H_2O$

(ii) $2H_2C = CH - CH_3 + 3NO \xrightarrow[Ag]{700°C} 2H_2C = CH - CN + \tfrac{1}{2}N_2 + 3H_2O$

Fig. 3.48 Commercial routes to acrylonitrile (cyanoethene) (a) from ethyne; (b) from ethene; (c) from propene (i) ammoxidation; (ii) from nitric oxide.

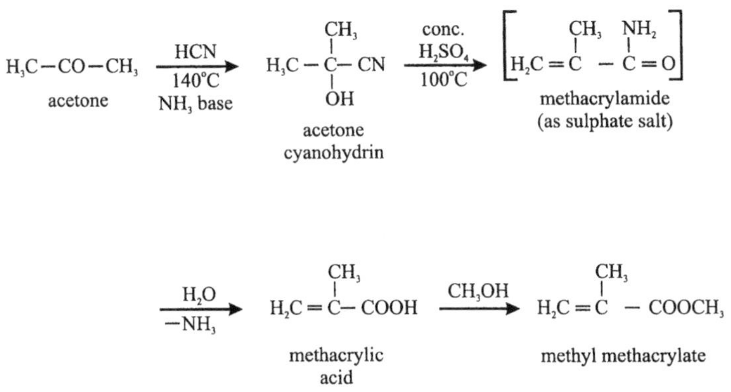

$\underset{\text{acetone}}{H_3C - CO - CH_3} \xrightarrow[\substack{140°C \\ NH_3\ base}]{HCN} \underset{\substack{\text{acetone} \\ \text{cyanohydrin}}}{\overset{CH_3}{\underset{OH}{H_3C - C - CN}}} \xrightarrow[100°C]{\substack{conc. \\ H_2SO_4}} \underset{\substack{\text{methacrylamide} \\ \text{(as sulphate salt)}}}{\left[\overset{CH_3\ \ NH_2}{H_2C = C\ -\ C = O} \right]}$

$\xrightarrow[-NH_3]{H_2O} \underset{\substack{\text{methacrylic} \\ \text{acid}}}{\overset{CH_3}{H_2C = C - COOH}} \xrightarrow{CH_3OH} \underset{\text{methyl methacrylate}}{\overset{CH_3}{H_2C = C\ - COOCH_3}}$

Fig. 3.49 Commercial synthesis of methyl methacrylate from acetone.

A newer route, avoiding toxic hydrogen cyanide, is a two-step catalytic oxidation of isobutene with air which converts $-CH_3$ to $-COOH$.

Dienes require a different synthetic strategy. The simplest diene, butadiene, is generally obtained as one of the products from the thermal cracking of

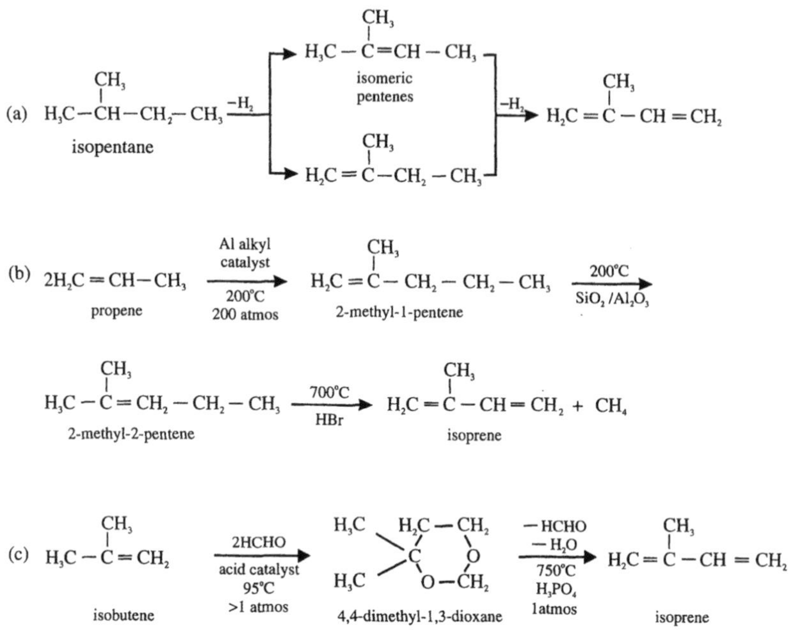

Fig. 3.50 Synthetic routes to isoprene (2 methyl butadiene) (a) thermal dehydrogenation; (b) propene process; (c) isobutene process.

petroleum. Other routes are based on dehydrogenation reactions, either from butanes or from other petroleum extracts.

Isoprene is a very important monomer whose synthesis was crucial to the development of artificial rubber. It can be prepared by a dehydrogenation route from isopentane (Fig. 3.50a) or by bond rearrangement and high temperature cracking reactions of alkenes, reactions that would not be considered in the small scale laboratory (Figs 3.50b and c). These are further examples of the importance of catalysis in industrial synthesis.

A variant on isoprene is chloroprene. This has a chlorine atom instead of the methyl group and requires a different synthetic strategy from butadiene. It polymerizes to give polychloroprene, also known as 'Neoprene', and was particularly useful in the Second World War as an artificial rubber with good weathering properties and resistance to heat and oil.

Fluorinated monomers

Since fluorine compounds are relatively expensive, it limits wider usage of the polymers. Also, elemental fluorine is an extremely corrosive, highly reactive and hazardous material. Fluorine chemistry is sufficiently distinct from conventional halo-organic chemistry for different strategies to be required and generally the most cost-effective procedure is to start from an appropriate chlorocompound. *Tetrafluoroethene* (PTFE) is prepared by a two-step process (Fig. 3.51). Because of its properties (high melting and insoluble), the polymer is very difficult to process and for some applications a copolymer with ethene or some other hydrocarbon system is preferred. Non-stick frying pans employ

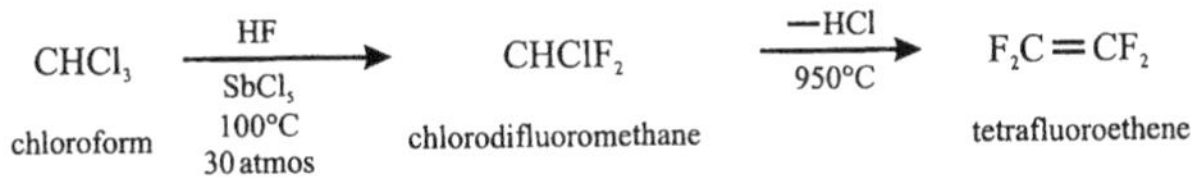

CHCl₃ $\xrightarrow[\substack{\text{SbCl}_5 \\ 100°C \\ 30\,\text{atmos}}]{\text{HF}}$ CHClF₂ $\xrightarrow[950°C]{-\text{HCl}}$ F₂C=CF₂

chloroform   chlorodifluoromethane   tetrafluoroethene

Fig. 3.51 Synthesis of tetrafluoroethene.

a series of coatings, the bottom-most being a copolymer that can stick to the metal, with the top-most being PTFE itself.

3.7 Properties of chain-growth polymers

The bulk polymers polyethene, polypropene, polystyrene, polyacrylic derivatives and polydienes are used on such a large scale and in so many applications that it is difficult to sum up specific properties within the constraints of this *Primer*, particularly since continued development and improvement has resulted in widespread usage as copolymers, blends, composites or after post-polymerization modification. However, it is hoped that a flavour will be imparted and the interested reader is referred to more specialist treatments (see Chapter 7) for particular areas of interest.

Even an apparently simple polymer such as *polyethene* has complexities in its various forms: low-density (LDPE), high-density (HDPE), ultra high molecular weight (UHMWPE) as well as copolymers such as linear high-density (LHDPE) (a copolymer with a substituted alkene to limit branching), EPM/EDPM rubbers (copolymers with propene and dienes), EVA (a copolymer with vinyl acetate), 'ionomers' (copolymers with low amounts of methacrylic acid in which divalent metals such as Mg^{2+} form ionic cross-links between chains to give strength), while sulphonated EPDM is an elastomeric ionomer because of the pendant ionic sulphonate groups.

Polystyrene is relatively brittle, but gives strength to copolymers. For example, where butadiene provides the elastomeric 'soft' block of the widely used styrene–butadiene rubbers (SBR). Acrylonitrile–butadiene–styrene (ABS) terpolymer further exploits the crystallinity and good fibre-forming capability of acrylonitrile. Acrylonitrile and butadiene themselves make an elastomeric copolymer, nitrile rubber (NBR) which is useful because of its outstanding resistance to petroleum and similar solvents.

Polyacrylonitrile (PAN) is used to produce 'carbon fibres', developed in the 1960s, where sequential pyrolysis (Fig. 3.52) produces a ladder polymer which is not 'all-carbon' as might be implied from the name.

Most commercial polymer systems are mixtures of components. For example, since pure PVC is relatively rigid ($Tg \sim 85$ °C), it is usually found plasticized with a heavy oil such as a phthalate ester. This can cause in-service problems in food packaging, where the wrapping of fatty foods (e.g. cheese) or the use of microwave heating can cause leaching out of the plasticizer.

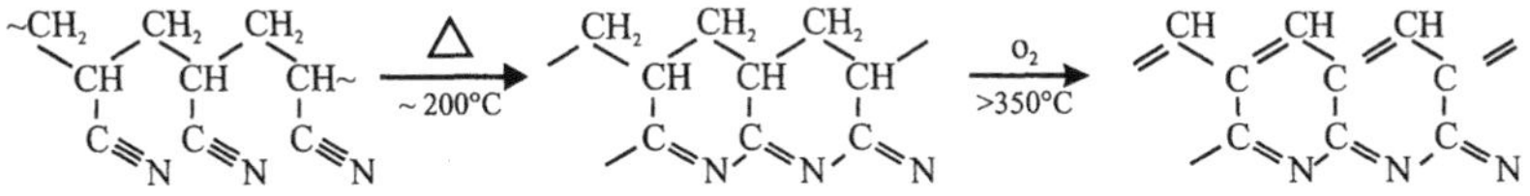

Fig. 3.52 Formation of carbon fibres from heating of polyacrylonitrile.

PVC copolymers with low amounts of vinyl acetate or vinylidene chloride have improved processing properties.

Polyacrylates give large scale glassy thermoplastics with considerable resistance to weathering. Soluble acrylic coatings can be thermoset with urethanes or epoxies to give tough insoluble networks widely used in the automobile industry.

Polyvinyl alcohol, polyvinyl pyrrolidone (PVP) and *polyacrylamide* are water-soluble polymers which increase solution viscosity, as do all dissolved polymers, and hence are used as thickeners and sizers for many applications including foodstuffs, cosmetics, pharmaceuticals, and adhesive pastes. PVP has a pronounced effect upon hydrophobic interactions in water and is used for the stripping of dyes from textiles. It was once used in a blood substitute. Polyacrylamide may also be cross-linked by a small proportion of a difunctional monomer ('bis acrylamide') to give a gel used in biology, where polyacrylamide gel electrophoresis (PAGE) is used to separate proteinaceous species on the basis of charge. A fresh gel is prepared *in situ* for each separation.

Rubber (polyisoprene) is still used on a large scale and a number of other post-treatments have been developed, apart from vulcanization, not all of which are intended mainly to produce cross-links. Thus 'chlorinated rubber' is not polychloroprene (2 chloro 1,3 butadiene) but is instead the product from treating rubber with chlorine gas in hot carbon tetrachloride. These free radical conditions also cause substantial substitution of hydrogen by chlorine to give polychlorination, with further complex chemistry. Other post-treatments of natural rubber include oxidation, reactions with sulphuric acid ('cyclized rubber') and hydrochloric acid. Chemical post-treatment is also used for other polymers and 'chlorinated PVC' has similarly improved properties.

The cross-linking 'vulcanization' reaction of polyisoprene (see Chapter 5) is largely responsible for its toughness while maintaining sufficient elasticity. However, there are other requirements for rubber goods. For instance car tyres must operate equally on cold winter mornings and hot summer afternoons. Furthermore, there is additional heat build-up during flexing under motion. Butadiene, without the space-disrupting methyl group of isoprene, gives a more crystalline polymer. Hence a modern tyre is a blend of various diene rubbers, including styrene–butadiene copolymer systems, chosen to manipulate Tg, Tm and other phase changes that influence temperature-dependent phenomena. The filler material, carbon black, is an important component and the development of the car tyre is itself a fascinating story of applied materials chemistry.

Fluorinated polymers are useful because of the extreme electronegativity and small size of the fluorine atom. The best-known, *PTFE*, has exceptional chemical and thermal stability. *Polyvinylidene fluoride (PVDF)* exhibits the uncommon phenomena of piezoelectricity and pyroelectricity which are the respective conversions of mechanical movement or temperature changes into electrical signals. This behaviour originates from the alternation of CH_2 and highly polar CF_2 units along the chain, these being forced into the same stereochemical attitude throughout the chain by the demands of the sp^3 carbon atoms. An alternating copolymer of $CH_2=CH_2$ and $CF_2=CF_2$ would not show the same effect. The polymer must be 'poled', that is, heated above Tg and oriented in an electric field before quenching to 'freeze' in the electronic effects. The opportunities for scale-up offered by polymers such as PVDF is of benefit in many functional applications (see Chapter 6).

Other fluorinated chain systems are based on hexafluoropropene ($CF_3CF=CF_2$) and perfluoromethyl vinyl ether ($CF_2=CF–O–CF_3$) and are used in copolymers with vinylidene fluoride and sometimes tetrafluoroethene to give modern high performance elastomers with improved chemical and thermal stability for use in gaskets, seals and other applications where high cost is acceptable. Such materials have various tradenames such as Viton™, Fluorel™, Kalrez™, and others.

Another high technology application of related materials is in selective ion-exchange membranes, exemplified by Nafion® from Du Pont. This is a perfluorinated PTFE/polyvinyl ether copolymer with the additional feature of a pendant sulphonic acid group as shown in Fig. 3.53. This copolymer was developed for the chlor–alkali industry (i.e. the electrolysis of sodium chloride in water) to keep the nascent chlorine (produced at the anode) separated from the sodium hydroxide (produced at the cathode) yet still allow the passage of ions.

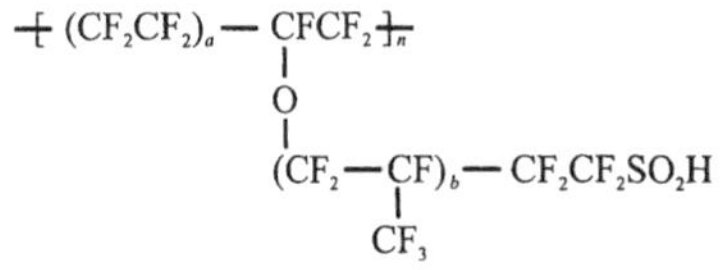

A general treatment of permeability through polymer membranes is beyond the scope of this book, but considerable developments have been made not just in the separations of ions, but also liquids and gases, and is of benefit to many applications. For example, the low permeability of polyethylene terephthalate to carbon dioxide prolongs the shelf-life of carbonated drinks in PET bottles, while the low general gas permeability of butyl rubbers (IIR) (isobutene copolymer with a little isoprene) results in use as inner tubes for tyres. Permeability and separation effects are not due just to the basic chemical structure of a polymer, but also due to pore size, bulk density and other physical effects which originate from preparation procedure and subsequent treatment. Polymers with pores ranging from 10 Å through to 10 μm can be produced (known as ultrafilters and microfilters) which as films are able to discriminate between particles of different sizes, including bacteria (for 'cold sterilization'), viruses, starch, proteins and even smaller molecules.

Porosity and permeability are important parameters in many other applications. Hard contact lenses are made from relatively impermeable polymethacrylates, while soft contact lenses employ hydroxyethyl methacrylate (HEMA) systems, in which the hydroxy-group imparts water-compatibility. The further development of modern contact lenses that can be left in place in the eye for long periods has been made possible by polymers with oxygen permeability that more closely matches the natural properties of the eye. This involves hybrid polymers with hydrophilic (polyvinyl pyrrolidone) and hydrophobic (polysiloxane) components.

Other modern developments include interpenetrating polymer networks (IPNs) in which a second polymer is polymerized separately inside a first polymer, usually by swelling the first polymer with the second monomer in solution and then initiation of polymerization. For example, acrylates can be polymerized inside a polyurethane matrix. IPNs can have different behaviours to either traditional copolymers made from the same components, or to the type of network which would be produced by subsequent cross-linking of acrylate units in an unsaturated polyurethane chain. IPNs show yet again how polymer properties can be varied by careful control of structure.

$$\begin{array}{c} \text{---}(CF_2CF_2)_a\text{---}\underset{\displaystyle \underset{\displaystyle (CF_2\text{---}CF)_b\text{---}CF_2CF_2SO_3H}{\overset{\displaystyle O}{|}}}{\overset{\displaystyle |}{\underset{\displaystyle |}{\overset{\displaystyle CFCF_2}{}}}}\text{---}_n \\ \overset{\displaystyle |}{CF_3} \end{array}$$

Fig. 3.53 Nafion® perfluorinated ionomer (in acid form).

4 Step-growth polymers

4.1 Introduction

In the case of step-growth (condensation) polymers, the mechanism is simply an extension of the normal organic condensation reactions in which a small molecule e.g. H_2O or HCl is expelled as the link is built. For example:

$$
\underset{\text{acid}}{CH_3\overset{\overset{\textstyle O}{\|}}{C} - CH} + \underset{\text{alcohol}}{HOCH_3} \longrightarrow \underset{\text{ester}}{CH_3\overset{\overset{\textstyle O}{\|}}{C} - O - CH_3} + H_2O
$$

or
$$
R - CO - Cl + H - O - R_1 \longrightarrow R - CO - OR_1 + HCl
$$

Also
$$
\underset{\text{amine}}{R - NH - H} + \underset{\text{acid}}{HO - OCR_1} \longrightarrow \underset{\text{amide}}{R - NH - OCR_1} + H_2O
$$

This is a different situation to the chain polymerizations described in the last chapter and the main differences between step and chain are given in Table 4.1.

Table 4.1 Comparison of step and chain polymerization

Step	Chain
Growth throughout matrix	Growth by addition of monomer only at one end of chain
Rapid loss of monomer species*	Some monomer remains even at long reaction times
Same mechanism throughout	Different mechanisms operate at different stages of reaction
Molar mass increases slowly throughout	Molar mass of backbone chain increases rapidly
Ends remain active (no termination)	Chains not active after termination
Polymerization rate decreases as the number of functional groups decreases	Polymerization rate increases initially then becomes constant
No initiator necessary[†]	Initiator required

*Monomer is consumed rapidly for step reactions since the first reaction is to produce dimer.
[†]Catalysts will of course assist reaction.

4.2 Kinetics of step polymerization

It is assumed that most step polymerizations involve bimolecular reactions as key mechanistic processes. Typical examples of the production of polyesters are given in eqns 4.1 and 4.2. (For polyamides replace OH by NH_2, see Chapter 1.)

$$HO-R-OH + HOOC-R-COOH \rightarrow HO[-\overset{\overset{O}{||}}{C}-R-\overset{\overset{O}{||}}{C}-O]_nH + H_2 \quad (4.1)$$

$$HO-R-COOH \rightarrow HO[R-\overset{\overset{O}{||}}{C}-O]_nH + H_2 \quad (4.2)$$

In general, the reactions given in eqns 4.1 and 4.2 may be represented by the general chemical equation (eqn 4.3) and the general kinetic equation (eqn 4.4).

$$-A + B- + \text{catalyst} \rightarrow -AB- + \text{catalyst} \quad (4.3)$$

$$\text{Rate} = k[A][B][\text{catalyst}] \quad (4.4)$$

If the esterification reaction is acid-catalysed (i.e. H^+) [catalyst] is usually taken to be $[H^+]$ and two mechanisms should be considered since H^+ ions can be supplied either from an external source (external catalysis) or provided internally by the dissociation of the carboxylic acid (self-catalysis).

If stoichiometry of the functional groups A and B is assumed, the kinetic equation (eqn 4.4) may be rewritten as eqn 4.5.

$$\text{Rate} = -d[A]/dt = k[H^+][A]^2 \quad (4.5)$$

It is also possible, by monitoring the loss of the acid groups, A, e.g. by titration with base, to estimate the concentration (or number of molecules) of carboxylic acid present initially and at some time t and hence to determine the degree of polymerization $\overline{X_n}$ (eqn 4.6) and make an estimate of the number average molar mass, $\overline{M_n}$ (eqn 4.7).

$$\overline{X_n} = \frac{N_o}{N_t} \quad (4.6)$$

$$\overline{M_n} = \overline{M_o} \times \overline{X_n} \quad (4.7)$$

In eqn 4.6, N_o represents the number of carboxylic groups present initially and N_t those present at time t. In eqn 4.7, $\overline{M_o}$ is the value of the mean molar mass and is given by eqn 4.8.

$$\overline{M_o} = \frac{\text{Molar mass of the repeat unit}}{\text{Number of monomer units in the repeat unit}} \quad (4.8)$$

Also, assuming equal numbers of the functional groups A and B (i.e. stoichiometry) allows a deduction as to how far the step reaction has proceeded towards producing polymer (i.e. extent of reaction, p) in a given period of time (eqn 4.9).

$$p = \frac{\text{Number of carboxylic acid groups which have reacted}}{\text{Number of carboxylic acid groups initially present}} \quad (4.9)$$

In terms of the number of carboxylic groups, eqn 4.9 can be rewritten as eqn 4.10.

$$p = \frac{(N_o - N_t)}{N_o} \tag{4.10}$$

Combining eqns 4.6 and 4.10 gives eqn 4.11.

$$\overline{X_n} = \frac{1}{(1-p)} \tag{4.11}$$

If a slight stoichiometric imbalance of the functional groups occurs, however, then the attainable molar mass (RMM) is significantly limited. For example, when the number of functional groups actually present is N_A and N_B, then $\overline{X_n}$ takes the value given in eqn 4.12 where r is known as the reactant ratio and is equal to $\frac{N_A}{N_B}$

$$\overline{X_n} = \frac{(1+r)}{(1+r-2rp)} \tag{4.12}$$

Equation 4.12 can also be used for reactions in which a monofunctional substrate (e.g. $CH_3\,CO_2H$ or CH_3OH) has been used to control the degree of polymerization. In this case r is defined by eqn 4.13 where N_{BM} is the number of monofunctional molecules, the multiplier two being required since one monofunctional group has only half the effect of a difunctional group (N_B).

$$r = \frac{N_A}{N_B + 2N_{BM}} \tag{4.13}$$

In general, for a reaction which undergoes external catalysis, the relationship between the extent of reaction (p) or number average molar mass $\overline{M_n}$ and time (t) is given by eqns 4.14 and 4.15, respectively.

$$\frac{1}{(1-p)} - 1 = k'[A]_0 t \tag{4.14}$$

$$\overline{Mn} = M_o(1 + k'[A]_0 t) \tag{4.15}$$

For a reaction which is self-catalysed, the extent of reaction and molar mass are given by eqns 4.16 and 4.17, respectively.

$$\frac{1}{(1-p)^2} - 1 = 2k[A]_0^2 t \tag{4.16}$$

$$\overline{M_n^2} = \overline{M_o^2}(1 + 2k[A]_0^2 t) \tag{4.17}$$

Note the altered dependence of RMM and build-up for self-catalysis. This must be borne in mind when designing a practical process.

4.3 Commercial preparations of step-growth polymers

Linear saturated polyesters, polyethylene terephthalate

This is the archetypal high-tonnage polyester first produced commercially licenced to ICI in the 1950s and used as fibre in textiles (tradename 'Terylene'), in film form as backing for photographic film, audio and video cassette tapes, and as moulded articles, e.g. bottles for carbonated drinks (useful for its low permeability towards carbon dioxide which avoids the carbonated drink going 'flat' in storage and prolongs product shelf-life).

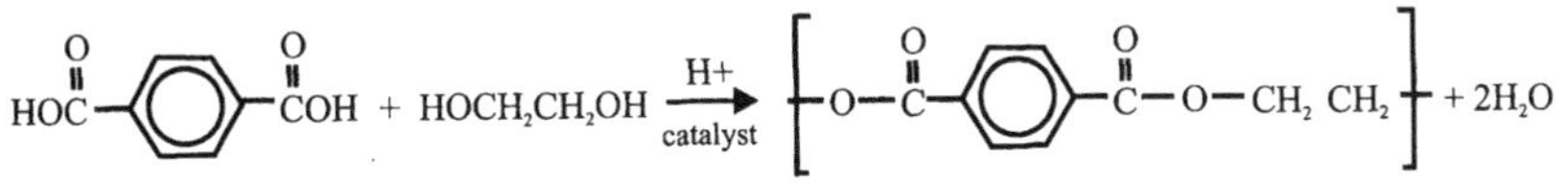

Fig. 4.1 Laboratory synthesis of poly(ethylene terephthalate).

Synthesis

To prepare polyethylene terephthalate (PET) on a laboratory scale requires the reaction under reflux of ethylene gylcol and terephthalic acid, with a trace of p-toluene sulphonic acid as catalyst using Dien–Stark apparatus to remove water as it is formed (i.e. to drive the equilibrium towards the ester) (Fig. 4.1).

However, a typical laboratory yield as high as 95% is still not sufficient for the production of high RMM step-growth polymers. A further complication is that before use the acid must be to be purified, which is difficult because of its high melting point and poor solubility characteristics. Thus a terephthalate derivative is necessary, such as the diacyl chloride, but this is significantly more expensive as a raw material and cannot be justified for a high-volume low-cost application.

The industrial preparation of PET therefore exploits the reversibility of esterification. The general principle is given in Fig. 4.2a–c and for the specific case of a methyl ester (Fig. 4.2d), where it can be seen that, as the original ester hydrolyses, the volatile component distils off to be permanently lost from the system. Re-esterification takes place with the involatile replacement component until the equilibrium reaches the product ester. Methanol is a useful volatile alcohol and has the benefit that virtually complete removal can be achieved. Water is more difficult to remove completely.

The full route to PET from the various raw materials is given in Figs 4.3–4.5. Note the use of catalysts, high and low pressures and heat which are straightforward to manipulate in industrial scale reactors.

For a typical commercial polymerization dimethylterephthalate is mixed with ethylene glycol in a ratio of 1:2.1. The slight imbalance is designed to

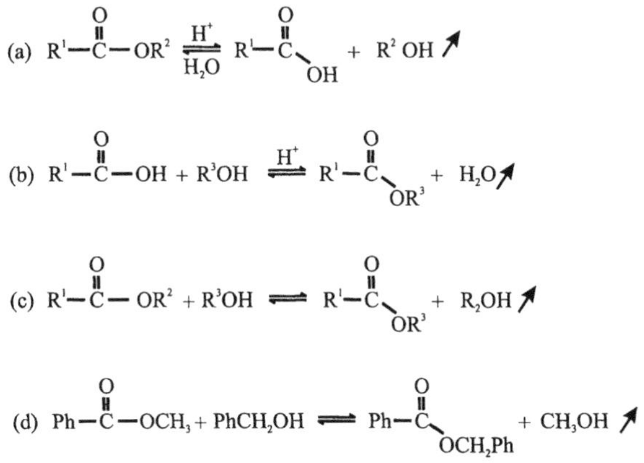

Fig. 4.2 Transesterification (Ester-Interchange) reactions (a) hydrolysis of first ester to evolve volatile alcohol; (b) re-esterification with an involatile alcohol; (c) overall process; (d) specific example.

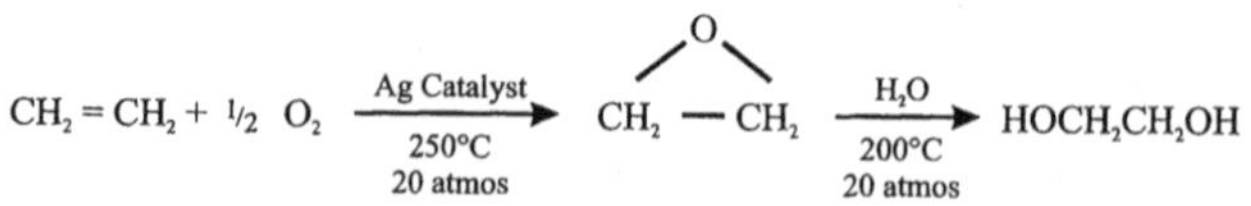

Fig. 4.3   Commercial synthesis of ethylene glycol (ethane 1,2 diol).

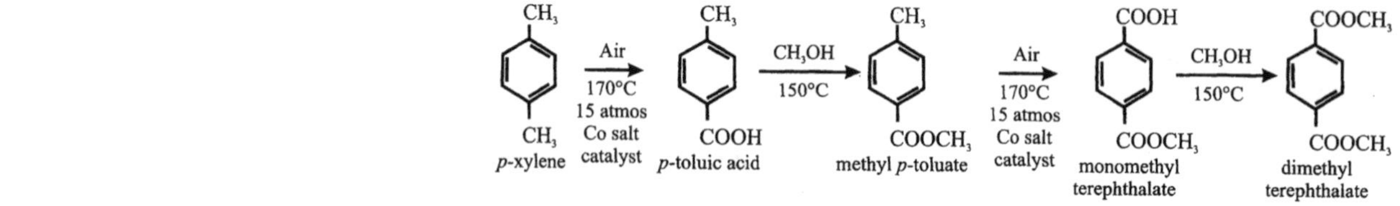

Fig. 4.4   Commercial synthesis of dimethyl terephthalate.

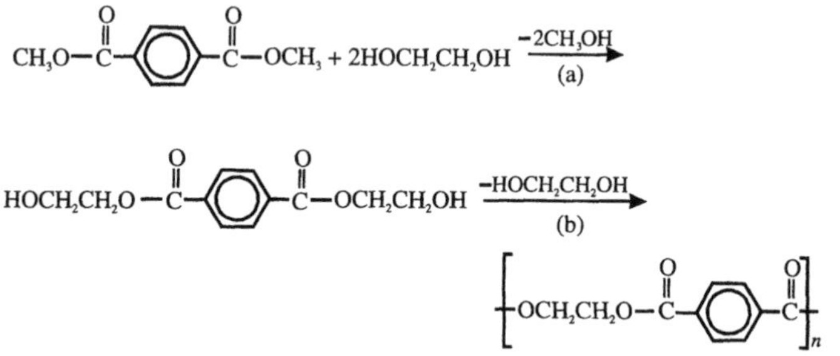

Fig. 4.5   Commercial syntheses of polyethylene terephthalate ('terylene' PET) (a) first stage (1 atm/ 200 °C); (b) second stage (10^{-3} atm/290 °C).

limit RMM (this can also be achieved by monofunctional chain stoppers), but note the use of two diol molecules per diester molecule in the starting mixture.

The polymerization takes place in two stages, as shown in Fig. 4.5. First, heating to some 200 °C at atmospheric pressure induces transesterification of methanol by ethylene glycol, together with some oligomerization and chain growth. In the second step the mixture is further heated to 290 °C under reduced pressure ($\sim$1 mm Hg). A second transesterification now occurs, in which one molecule of ethylene glycol is released, while the other remains to react with a nearby molecule to build up the polymer chain.

The economics of any chemical process depend upon recycling where possible and here methanol and the recovered ethylene glycol are re-used.

Both steps employ catalysts, typically manganese acetate (or another equivalent metal salt) for the first step while antimony trioxide is now favoured for the second.

Modern developments in purification now allow direct use of terephthalic acid in the polymerization reaction. This gives the same intermediate after the first step and the second step is the same.

In either case the final reaction continues to the desired RMM, 20 000 for thin film applications and 30 000 for moulding.

Processing of polyethylene terephthalate

A key feature in polymer technology is the manipulation of polymer properties by post-polymerization processing. PET behaviour is affected by its crystallinity (melting temperature T_m is 265 °C). Typically for fibres the molten as-prepared polymer is extended through spinnerets into ambient air, cooling to an amorphous material, which is permanently elongated under stress ('drawn') above the glass transition temperature (T_g) of 80 °C to induce orientation; it is then drawn again under tension at 200 °C to induce maximum crystallinity. Similar multiple processes are used for films and for moulding.

Typical tensile strengths of PET in various forms are: Fibres ~700 MPa, film ~180 MPa, amorphous moulding 55 MPa and crystalline moulding 76 MPa. PET is a good electrical insulator. It is chemically resistant to water, dilute (but not strong) mineral acids and bases, but organic bases diffuse through and attack the bulk so contact with amines is undesirable. To some extent PET is being replaced by polybutylene terephthalate (PBT) in which $HO-(CH_2)_4-OH$ replaces $HO-(CH_2)_2-OH$ as the diol component. The reason is not because of polymerization phenomena, but is instead due to the demands of processing protocol. PET takes time to crystallize during processing while PBT crystallizes more rapidly. This avoidance of delay is sufficient improvement to warrant the alternative polymer.

Polyester plasticizers

Simple esters such as dioctyl phthalate (diethylhexyl phthalate) are widely used as plasticizers in the polymer industry, for example to modify the properties of polyvinyl chloride (PVC) (Chapter 3), but it is often desirable to use materials with somewhat higher RMM which are less volatile and more difficult to leach out. This is important in microwave applications. Polyesters of moderate RMM (500–10 000) are useful and the diols used include ethane, propane, butane derivatives, including polyetherols such as low RMM polyethylene glycol, while diacids are usually aliphatic rather than aromatic e.g. adipic acid and sebacic acid. Such dihydroxy-ended resins are used as segments of other polymer systems such as polyurethanes (See later). Mono-ols or mono-acids may be used as end-cappers for control of RMM.

Linear saturated polyamides (nylons)

These were the great success of Carothers at Du Pont de Nemours (USA), who started a study into industrial polymerization in 1929. Carothers also investigated polyesters but used aliphatic materials rather than the aromatic units in the later successful PET. These aliphatic polyesters had low melting points and poor resistance to hydrolysis. However, nylons, which are substantially hydrogen-bonded in the solid state via the N–H units in the amide linkages, produce effective materials even with wholly aliphatic groups in the backbone.

The commercial availability of nylon 6,6 in 1940 for nylon stockings led to immediate success because of the high added value application for hosiery in the fashion industry. The search for nylon analogues not covered by Du Pont's Patents led to the development of the isomeric polymer nylon 6 by I. G. Farben Industrie in Germany in 1940. Structures of the most

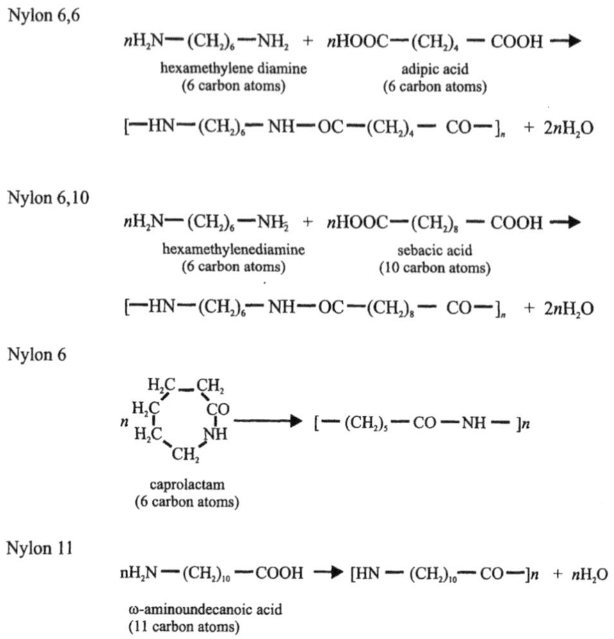

Nylon 6,6

$$nH_2N—(CH_2)_6—NH_2 \ + \ nHOOC—(CH_2)_4 — COOH \longrightarrow$$

hexamethylene diamine   adipic acid
(6 carbon atoms)   (6 carbon atoms)

$$[—HN—(CH_2)_6— NH—OC—(CH_2)_4— CO—]_n \ + \ 2nH_2O$$

Nylon 6,10

$$nH_2N—(CH_2)_6—NH_2 \ + \ nHOOC—(CH_2)_8 — COOH \longrightarrow$$

hexamethylenediamine   sebacic acid
(6 carbon atoms)   (10 carbon atoms)

$$[—HN—(CH_2)_6— NH—OC—(CH_2)_8— CO—]_n \ + \ 2nH_2O$$

Nylon 6

$$\longrightarrow [—(CH_2)_5—CO—NH—]n$$

caprolactam
(6 carbon atoms)

Nylon 11

$$nH_2N—(CH_2)_{10}—COOH \longrightarrow [HN—(CH_2)_{10}—CO—]n \ + \ nH_2O$$

ω-aminoundecanoic acid
(11 carbon atoms)

Fig. 4.6 Structures of important nylon polyamides.

commercially important nylons and the nomenclature system are given in Fig. 4.6. Note the differential directional senses of the linking groups nylon 6,6 and nylon 6 which are isomers. This has subtle effects upon polymer properties (see page 6). Nylon 6,6 and nylon 6 are the most widely used examples, but a range of other nylons are based on the two isomeric types of which nylon 6,10 and nylon 11 are also exemplified in Fig. 4.6.

Raw materials
Nylon 6,6
Both components are 6-carbon units and it is cost-effective to prepare one through the other. The cheapest 6-carbon feedstock is benzene which is converted to adipic acid by first hydrogenation and then oxidation in two different steps (Fig. 4.7).

Adipic acid may be converted to hexamethylene diamine via the dinitrile as shown in Fig. 4.8. Note again the use of high pressures and catalysts.

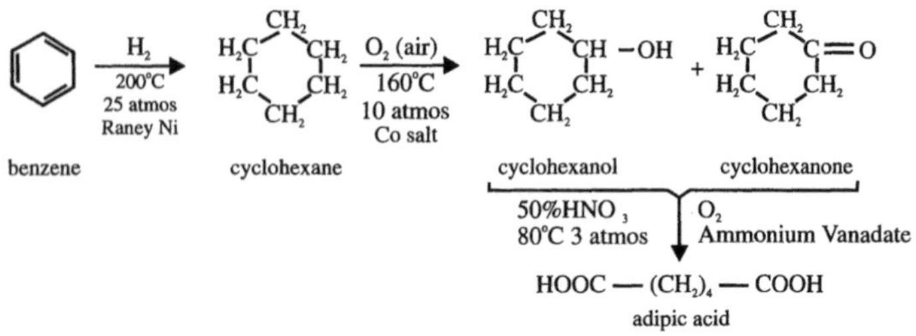

Fig. 4.7 Commercial route to adipic acid.

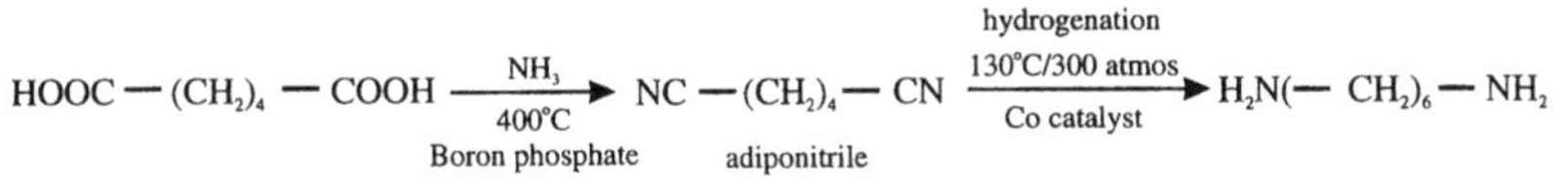
HOOC — (CH2)4 — COOH
NH3
400°C
Boron phosphate
NC — (CH2)4 — CN
adiponitrile
hydrogenation
130°C/300 atmos
Co catalyst
H2N(— CH2)6 — NH2

Fig. 4.8 Conversion of adipic acid to hexamethylene diamine.

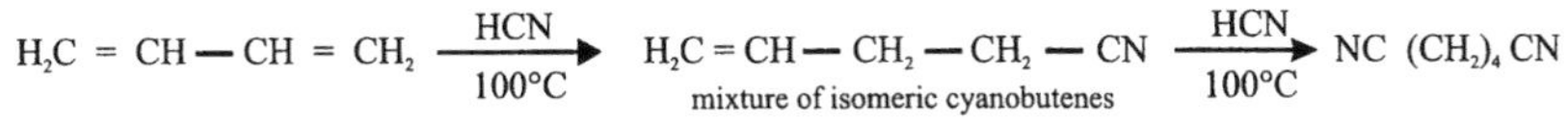
H2C = CH — CH = CH2
HCN
100°C
H2C = CH — CH2 — CH2 — CN
mixture of isomeric cyanobutenes
HCN
100°C
NC (CH2)4 CN

Fig. 4.9 Conversion of butadiene to adiponitrile.

Availability of starting materials is economically important to industry and here there is an opportunity to convert chain polymerization monomers into materials useful for step-growth polymerization. There are two recent routes from chain monomers to adiponitrile, which is a useful intermediate since it can be either hydrolysed to adipic acid or reduced to hexamethylene diamine to give both the necessary components for nylon 6,6.

Adiponitrile from butadiene This requires two successive additions of hydrogen cyanide (Fig. 4.9). The first step gives a mixture of isomers, which must be separated before continuation.

Electrochemical route to adiponitrile
This was developed by Monsanto for converting acrylonitrile to adiponitrile by electrochemical hydrodimerization (Fig. 4.10).

A number of side-products are possible and optimization has required substantial development. This process remains a high tonnage commercial electro-organic process now used by BASF.

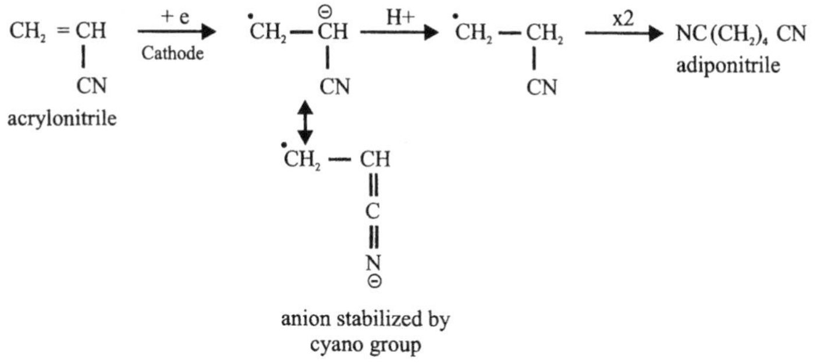
+e
Cathode
acrylonitrile
H+
x2
NC(CH2)4 CN
adiponitrile
anion stabilized by
cyano group

Fig. 4.10 Electroreductive route from acrylonitrile to adiponitrile.

Sebacic acid (HOOC $(CH_2)_8$ COOH
This may still be produced from the natural product castor oil which is an ester of glycerol. Hydrolysis gives ricinoleic acid, $(CH_3(CH_2)_5CH(OH)CH_2CH{=}CH{-}(CH_2)_7COOH$, which is oxidatively cleaved. This gives just a 50% yield from castor oil but remains viable because of the biorenewable starting material.

There is also an electrochemical route from adipic acid to sebacic acid, using the Kolbe electro-oxidation of carboxylate salts (Fig. 4.11). First adipic

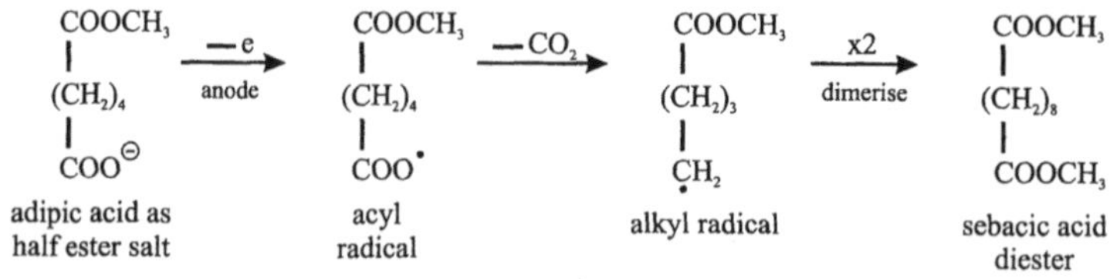

Fig. 4.11   Electro-oxidative route from adipic acid to sebacic acid derivative.

acid is converted to the half-ester to protect one of the carboxylate groups against electrochemical reaction. Then the half-ester carboxylate salt is electro-oxidized via an acyloxy radical, with loss of carbon dioxide, to an alkyl radical that dimerizes to the product.

The Kolbe reaction has a number of side-products and optimization has required careful control of conditions.

Typical Polymer preparation for Nylon 6,6 (also Nylon 6, 10 and similar polymers)

Careful control of stoichiometry is required to ensure high RMM and even for low RMM products it is desirable to have exact stoichiometry before adding monofunctional end-cappers to limit RMM. Similar considerations apply to polyesters and any step-growth polymers prepared from two difunctional molecules (i.e. X–X and Y–Y reacting together to give X–XY–YX–X, etc).

A school level demonstration of nylon 6,6 is simply to place immiscible diamine and diacyl components in a beaker and pull away a strand of polymer as it forms at the liquid–liquid interface, giving a very simple fibre-producing system to demonstrate the basic principle. However, the polymer so formed has poor properties and for commercial production the stoichiometry is nicely controlled by noting that amines are organic bases. Neutralization of the dicarboxylic acid with the diamine in boiling methanolic solution precipitates the 1:1 salt, which is filtered and dissolved in water for polymerization (Fig. 4.12). Alternatively, neutralization can take place in water, using conductimetric determination of the endpoint. Here the aqueous salt solution is used directly onwards.

In the polymerization of this or similar nylon variants a concentrated solution of the nylon salt, with a small amount of ethanoic acid (acetic acid) as cheap monofunctional end-capper, is heated to 220 °C giving some 20 atmospheres' pressure. After a few hours the temperature is raised to 280 °C, combined with venting to allow steam and volatiles to evolve, eventually leaving 1% moisture in the system. The hot melt is extruded under nitrogen since oxygen causes discolouration. (This consideration applies to many hot polymer melts.)

Nylon 6

Nylon 6 is used on a very large scale. The monomer caprolactam is the cyclic amide of ω-aminohexanoic acid (aminocaproic acid), which is a diheterofunctional species that automatically has the correct stoichiometry.

The polymer is made from the lactam rather than the amino acid since water does not need to be removed during the chain build-up. Furthermore, the lactam does not need to be prepared from the amino acid. Instead, strategies

Fig. 4.12   'Nylon salt' to give perfect stoichiometry for nylon 6,6.

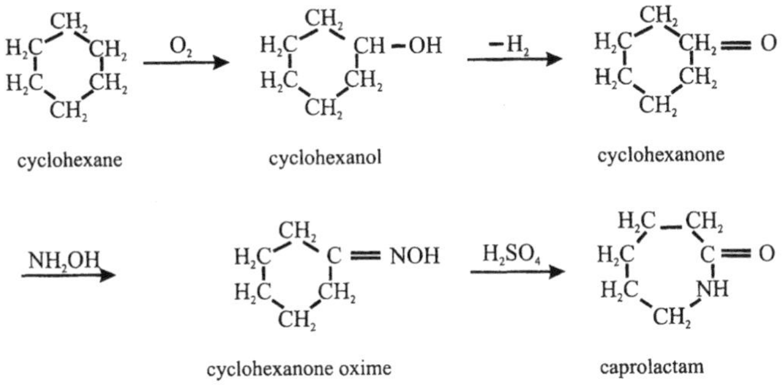

Fig. 4.13 Commercial route to ω-caprolactam.

from cheap starting materials directly to the cyclic molecule nicely demonstrate industrial approaches.

The best known lactam synthesis is shown in Fig. 4.13 and involves conversion of cheap cyclohexanone (from benzene (see Fig. 4.7) into the oxime, which undergoes an acid-catalysed Beckmann rearrangement directly to the lactam.

The oxime is isolated by alkaline neutralization with cheap aqueous ammonia to an oil which rearranges with oleum (SO_3/H_2SO_4) to give a product that requires neutralization again with ammonia. This generates large quantities of ammonium sulphate and the economics depend upon whether this byproduct can also be exploited.

An improved route, without isolation of the oxime and less acid (and hence less sulphate salt byproduct), uses amino sulphuric acid to take the ketone directly to the rearranged product without isolation of the oxime, and this saves overall on the process by removing a step.

Another interesting route uses photochemistry which, like electrochemistry, is not as widely used on a bulk industrial scale as might be expected. A different type of plant is needed compared to a thermal reaction. Cyclohexane and nitrosyl chloride in hydrochloric acid are irradiated by a mercury vapour lamp (Fig. 4.14). The reaction is cooled to −10 °C (a feature of photochemical

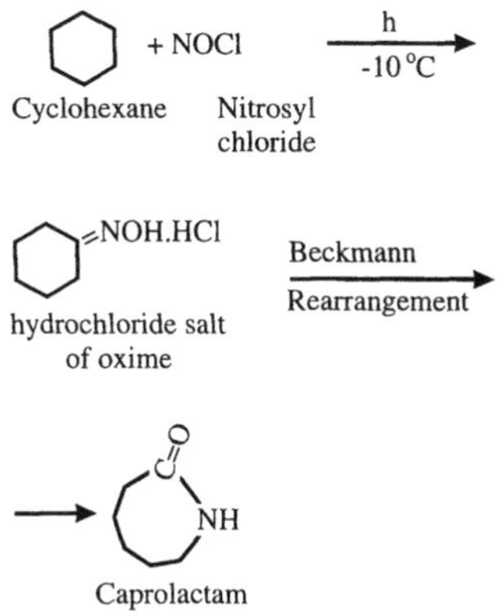

Fig. 4.14 Photochemical route to ω-caprolactam.

reactions is that true light initiation does not require heat which can cause side reactions). Overall the raw material costs are less and there is less salt side-product formed.

Polymerization to nylon 6
Caprolactam is mixed with 1% ethanoic (acetic) acid as end-capper for RMM control and 5–10% (by weight) of water, then heated to 200 °C for 12 h, venting pressure to keep at 15 atm. The molten product is processed onwards.

The water opens some of the lactam to the ω-amino acid, but polymerization also occurs by transamidation (amide interchange) in which the lactam is opened by an amino end-group without release of water.

Chemically, the lactam opening is initiated by a nucleophile and any suitable anionic species, lone pair species or otherwise electron-rich species could suffice. This is exploited in an anhydrous variant of the process which can be employed in reaction injection moulding (RIM), a process which was developed in Germany in the 1960s for polyurethanes in which polymerization occurs actually in the mould. The anhydrous version of the caprolactam route to nylon 6 also has suitable kinetics, provided a powerful anionic initiator such as an alkali metal, a metal amide or metal hydride is used. The chemistry is beyond the scope of this *primer*, but provides a nice example of how an understanding of kinetic and mechanistic factors may be put to commercial use.

Nylon copolymers
While a mixture of monomers leads to copolymers (e.g. of nylon 6,6 and 6, 10) it is more convenient to make these by simply heating a blend of preformed nylon polymers above the melting point, where transamidation reactions first put blocks of different polymers into each others' chains but, with time, lead to an essentially equilibrium random copolymer.

Properties of nylons
Simple polyamides of this type are tough, flexible, have high-impact strength and are resistant to abrasion. The length of aliphatic unit alters hydrogen bonding and affects glass transition and crystallization and hence mechanical and physical properties and processing characteristics. This is further altered by copolymerization and can be seen by comparison of several materials in Table 4.2. The significant water absorption of simple nylons mitigates against use in electrical insulation, despite their ease of processing such applications.

Other linear step-growth polymers
This is a large topic which includes polyurethanes, polycarbonates, polysiloxanes, polyethers, polyamides and related polymers and only brief

Table 4.2 Representative properties of commercial grades of nylons

	6,6	6	6,10	11	6,6:6,10 (1:2 copolymer)
Crystalline melting point (°C)	265	215	215	185	195
Tensile strength (MPa)	80	75	60	40	40
Elongation at break (%)	90	150	130	280	200
Impact strength, (J/m)	40	45	95	95	110
Water absorption at saturation (%)	8.0	9.0	2.5	2.0	6.5

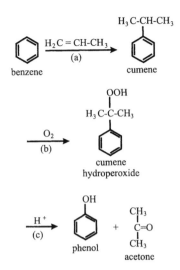

Fig. 4.15 Commercial production of phenol and acetone from benzene and propene (a) Friedel–Craft alkylation (with rearrangement); (b) aerial oxidation of cumene; (c) carbon–oxygen rearrangement and hydrolysis.

details can be given. Before describing the polymerizations it is worth mentioning 2, 2 bis(4′ hydroxyphenyl) propane, better known throughout the industry as bisphenol A, which is used for many step-growth polymers. This combines the rigidity and mechanical strength of the aromatic ring with the flexibility of the central tetrahedral sp^3 carbon. The space-filling methyl groups also protect against benzylic reactions that could occur if CH_2 was attached to the aromatic rings. The importance of this molecule is its easy synthesis from acetone and phenol, both of which are now relatively cheaply made in an efficient process starting from the petroleum product benzene (Fig. 4.15).

To make bisphenol A, excess phenol and acetone are mixed together and warmed to 50 °C while hydrogen chloride is passed through and the product precipitates. The mechanism is given in Fig. 4.16.

In effect bisphenol A is produced from cheap benzene and propene, without use of difficult conditions or complex and expensive reagents, and this is the origin of its widespread use in polymers. It may take part directly as a monomer diol, or else other functionalities such as amines or oxiranes (epoxides) may be inserted by nucleophilic substitution at the hydroxyl groups.

It is widely used in polycarbonates, which are tough impact-resistant polymers that have good dimensional stability, low combustibility and, importantly, transmit light well. This last quality derives from the effect of microcrystallite size upon light scattering and is a physical property rather than a chemical one. Another well recognized transparent polymer is the chain-growth perspex (polymethyl methacrylate). Polycarbonates are used as motorcycle helmet visors and as glass replacements in windows as well as other applications that make them the second largest volume thermoplastic. They are also the bulk polymeric material in compact discs. Where better elasticity is required they may be blended with acrylonitrile–butadiene–styrene (ABS) rubbers or similar materials. An ability to blend involves the

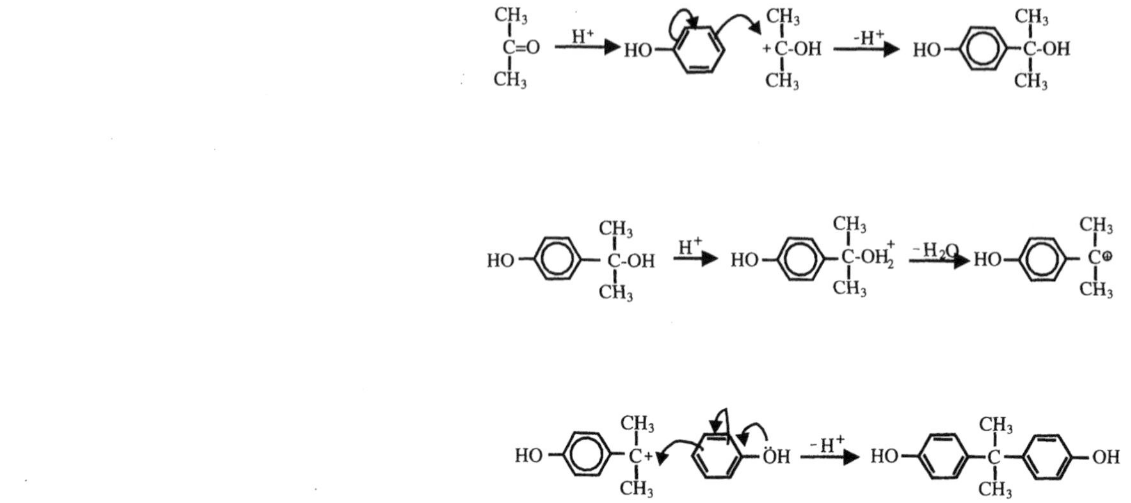

Fig. 4.16 Formation of bisphenol A by acid-catalysed mechanism.

thermodynamics of polymers and is a subject in its own right, but some basic principles are discussed in Chapter 2.

Ring-substituted bisphenol A (e.g. with methyl groups) is used to give altered properties, while halogenated bisphenol A confers flame retardancy, but monomer cost increases. However, currently there are some environmental concerns over large-scale use of Bisphenol A and its derivatives.

Polycarbonates

With bisphenol A, polycarbonate may either be formed by direct reaction with phosgene (carbonyl chloride) in alkaline solution, or else in pyridine (with a halogenated cosolvent for better economics) (Fig. 4.17a). Commercial viability of the latter process depends on recycling the expensive pyridine.

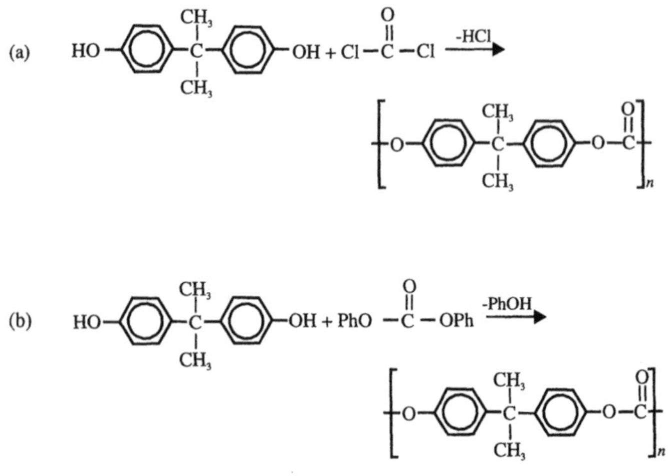

Fig. 4.17 Commercial syntheses of polycarbonates (a) from phosgene; (b) from diphenyl carbonate.

$$n \ HO{-}(CH_2)_4{-}OH \ + \ n \ O{=}C{=}N{-}(CH_2)_6{-}N{=}C{=}O \longrightarrow$$

1,4-butanediol   1,6-hexane di-isocyanate

$$\Big[\!{-}O{-}(CH_2)_4{-}O{-}\overset{\displaystyle O}{\overset{\|}{C}}{-}NH{-}(CH_2)_6{-}NH{-}\overset{\displaystyle O}{\overset{\|}{C}}\!\Big]_n$$

polyurethane

Fig. 4.18 Representative formation of a polyurethane.

Another route to polycarbonate employs the transesterification of diphenyl carbonate with the diol in the melt phase over a basic oxide catalyst. Phenol is evolved to drive over the equilibrium (Fig. 4.17b).

Despite the higher temperatures, vacuum apparatus and cost of diphenyl carbonate, the process is economic because it is free from solvents. The product can be used without purification and processing from the melt is straightforward.

Polycarbonates are very stable in the bulk, but the surface has a propensity to crack with stress (or craze), either with ageing, strain, or chemical exposure. This can cause in-service breakdown for certain applications.

Polyurethanes

These were first developed by Bayer in 1937 and usually involve reaction of a di-isocyanate with a diol. Strictly this is an addition reaction because no small molecule is expelled, but it follows a step-growth mechanism (Fig. 4.18).

The urethane linkage may be considered as a cross between a polyamide and a polyester, with good heat and chemical stability. They also have good abrasion resistance (and are used as soles for shoes), but can be somewhat susceptible to bacterial attack in the environment.

The commonest di-isocyanates in practical applications are given in Fig. 4.19.

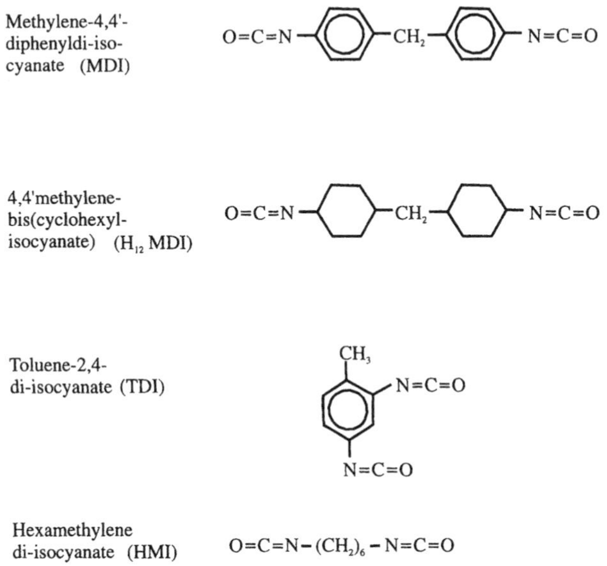

Methylene-4,4'-
diphenyldi-iso-
cyanate (MDI)

$$O{=}C{=}N\!-\!\!\bigcirc\!\!-\!CH_2\!-\!\!\bigcirc\!\!-\!N{=}C{=}O$$

4,4'methylene-
bis(cyclohexyl-
isocyanate) (H$_{12}$ MDI)

$$O{=}C{=}N\!-\!\!\bigcirc\!\!-\!CH_2\!-\!\!\bigcirc\!\!-\!N{=}C{=}O$$

Toluene-2,4-
di-isocyanate (TDI)

Hexamethylene
di-isocyanate (HMI)

$$O{=}C{=}N{-}(CH_2)_6{-}N{=}C{=}O$$

Fig. 4.19 Representative isocyanates used for commercial polyurethanes.

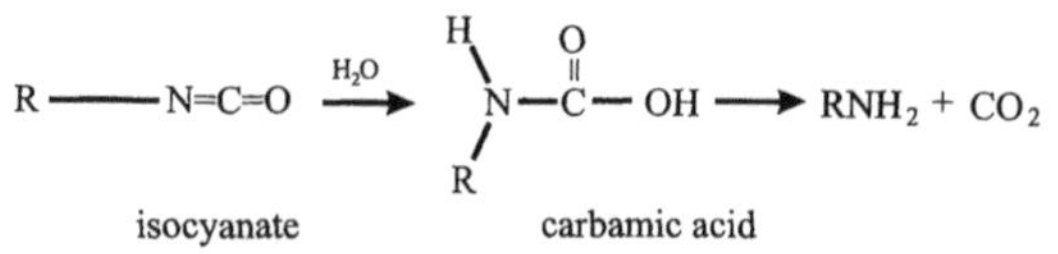

$$R \longrightarrow N{=}C{=}O \xrightarrow{\text{H}_2\text{O}} \begin{matrix} \text{H} & \text{O} \\ \diagdown & \parallel \\ \text{N}-\text{C}-\text{OH} \\ \diagup \\ \text{R} \end{matrix} \longrightarrow RNH_2 + CO_2$$

isocyanate   carbamic acid

Fig. 4.20   Reaction of water with an isocyanate to evolve carbon dioxide.

The dihydroxy function could be a simple diol, such as the ubiquitous bisphenol A for rigidity and toughness, but often polyurethanes employ hydroxy-ended short-chain oligomers, usually polyethers or polyesters, which are readily available.

Practical polyurethanes are thus effectively copolymers and, by choice of polymer structure, filler and other system components, they can be produced with a wide range of physical properties, from great strength (for bumpers, dashboards (fascias), panels and other uses in the automobile industry), through thermoplastic elastomers (rubber substitutes that can be melt-processed) to varnishes, lacquers and coatings. Unsaturated variants can be toughened by cross-linking (see fibreglass and paints, Chapter 5). The presence of water during polymerization produces space-filling foams since carbon dioxide evolves from the thermally unstable carbamic acid intermediate (Fig. 4.20). These foams may be either flexible or rigid, depending upon the components of the polyurethane system.

The kinetics of polyurethane formation are also suitable for Reaction Injection Moulding (RIM), which was invented during the commercial development of these polymers.

Polyurethanes are in principle very useful, but the di-isocyanate monomers are toxic, while the polymers unfortunately evolve hydrogen cyanide during combustion. Modern usage is becoming restricted on safety grounds; for example, foams are no longer employed in furnishings.

Polysiloxanes ('silicones')
Silicon is in the same column of the periodic table as carbon and is likewise tetravalent. However, the extra orbital shell alters bond angles and reactivities.

Silicon–hydrogen bonds are very reactive and silane (SiH_4) is pyrophoric (spontaneously inflammable in air). However, silicon–carbon bonds are relatively stable (tetramethylsilane, TMS, is the traditional internal standard for NMR spectroscopy), and silicon–halogen bonds are amenable to typical displacement reactions while silicon–oxygen is an extremely strong bond. Silicones are typically prepared from dihalodialkyl silanes either as linear or network polymers as shown in Fig. 4.21. The hydroxy-intermediates are not isolated and react onwards to the products.

Copolymers are also common, often formed by taking preformed cyclic derivatives and opening them up to rejoin linearly.

The chain is flexible, because of the wide SiOSi bond angle and because there is no substituent on the oxygen atom, which only carries the diffuse electron density of the lone pair.

Compare this with a carbon-based backbone, where there is a heavy nucleus (even if only a proton) on all adjacent atoms which are at the

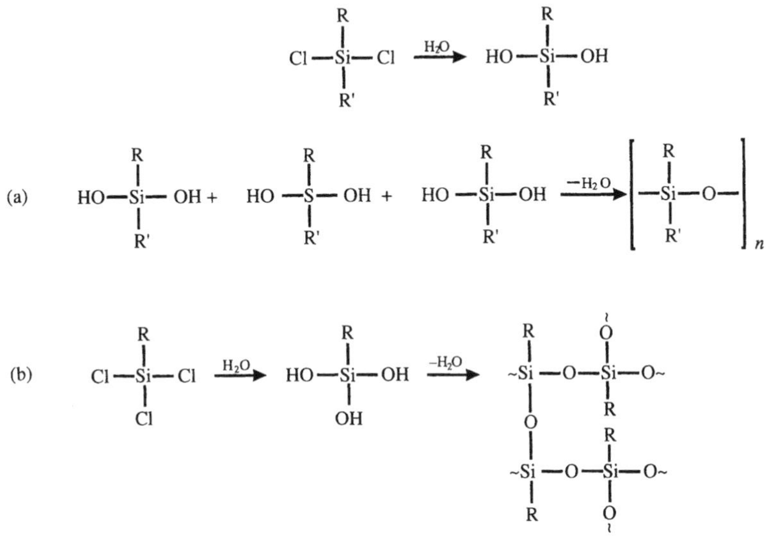

Fig. 4.21 Formation of polysiloxane from hydrolysis of haloalkylsilane (a) linear polymer; (b) network polymer.

tetrahedral bond angle of 109 °C. Carbon atoms are also smaller than silicon atoms with consequent crowding of attached groups.

Properties

Silicone elastomers are tough solids, particularly when cross-linked, but the same fundamental chemistry of chain formation extends down through intermediate materials to mobile liquids. Polysiloxanes are stable to oxygen in air towards 200 °C and they are stable to dilute acids and bases because of their ability to repel water. This is a very useful feature, exploited in the use of silicone fluids as polishes and as water-repellent sprays (e.g. for automobile electrical systems). They are good electrical insulators.

Polysiloxanes were laboratory curiosities until commercial production by Dow in the 1940s. They are now well known in technology but are still relatively costly.

A good example in which a small change in polymer structure gives interesting effects is 'silly putty' (bouncing putty). This is in essence a polydimethyl siloxane that has been heated with boric oxide in the presence of ferric chloride to give a low (< 1%) number of B–O–Si bonds in the structure. This material shows time-dependency of mechanical properties, such that if stretched slowly it is extremely elastic and likewise bounces if gently dropped; but if stretched rapidly with the same force it snaps. These unusual characteristics are thought to originate from polar interactions between lone pairs of oxygens in the chain and the electron-deficient boron atom. Borosilicone polymers are also used on a larger scale as 'fusible' elastomers.

Polyethers

By far the most widespread polyethers are cellulose derivatives based on the naturally occurring chain (Fig. 1.7), although the specific term 'cellulose

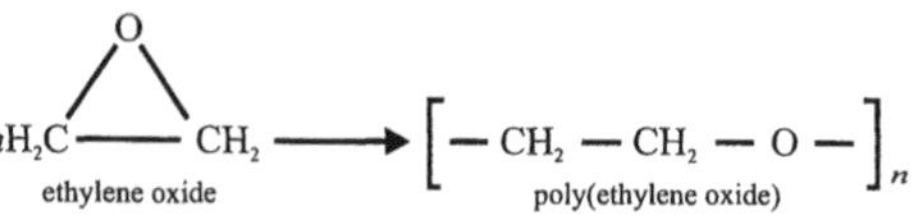

Fig. 4.23 Formation of polyethylene oxide.

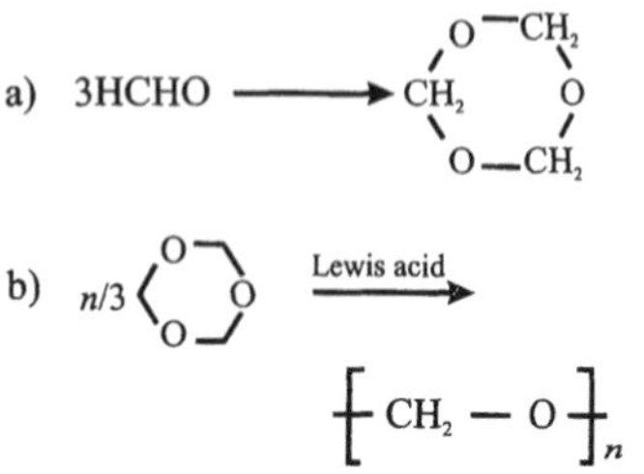

Fig. 4.22 (a) Trimerization of methanal to trioxan; (b) polymerization of trioxan to polyoxymethylene.

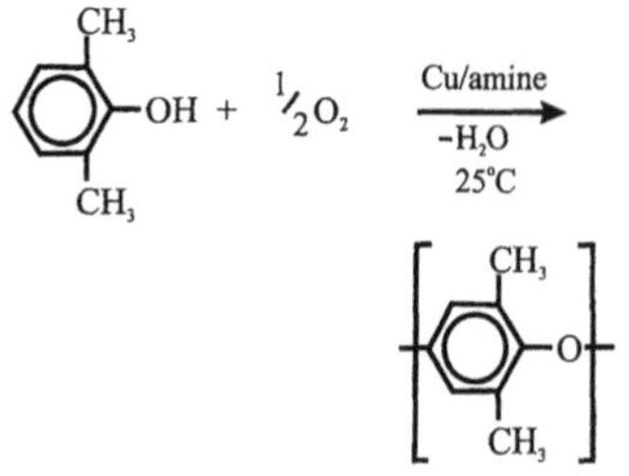

Fig. 4.24 Oxidative synthesis of an aromatic polyether.

ethers' refers to additional esterification at the pendant alcohol groups (see Chapter 1).

The largest scale artificial polyether is polyacetal (polyoxymethylene), which is essentially the polymer of methanal (formaldehyde), first produced in the late 1950s. Pure formaldehyde as the trimer trioxan polymerizes over a Lewis acid catalyst (Fig. 4.22), the important feature being the subsequent protection of the polymer end-groups by esterification with acetic anhydride.

Polymer RMM is up to 100 000 and, with such a simple structure, it is highly crystalline and tough. It is used as an engineering plastic for gearwheels and the like and has good dimensional stability and resistance to creep. It is also used in moderate temperature applications such as electric kettles. It is less stable to strong acids, bases and oxidizing agents and must be protected against air and ultraviolet light in service by antioxidant fillers.

Ethylene oxide (oxirane) may be opened by acid or base catalysts (Fig. 4.23) to give polyethyleneglycol (PEG) resins that are water-soluble and have useful non-ionic surfactant properties. They are also employed as thickening agents in a variety of applications, including food and hair shampoos. Control of conditions allows high RMM polymers which can have end-groups other than hydroxy groups. Polyethylene oxide (PEO) is highly crystalline, is readily processed (e.g. by calendering) and has a special feature in that it can act as a solid polymer electrolyte. This is an example of a functional property (see Chapter 6) and is demonstrated in the anhydrous polymer, where a salt such as lithium perchlorate displays ionic conductivity as though the solid polymer was a solvent. This offers possibilities of all solid-state batteries and other electrochemical devices.

Aromatic polyethers are typified by poly(phenylene oxide) (PPO), (Fig. 4.24). This polymer is highly resistant to aqueous acids and alkalis. It has good dimensional stability and a low coefficient of thermal expansion.

A further variant is poly(phenylene sulphide) (PPS), which is a thioether (Fig. 4.25). This is a strong stiff material, a good engineering plastic, chemically resistant with good thermal stability, maintaining mechanical properties to 250 °C. It has good flame retardency and a useful modern aspect is transparency to microwave radiation.

PPS variants are aromatic polysulphones. The simple equivalent of PPS, i.e. with SO_2 instead of S in Fig. 4.25, decomposes at its melting point beyond

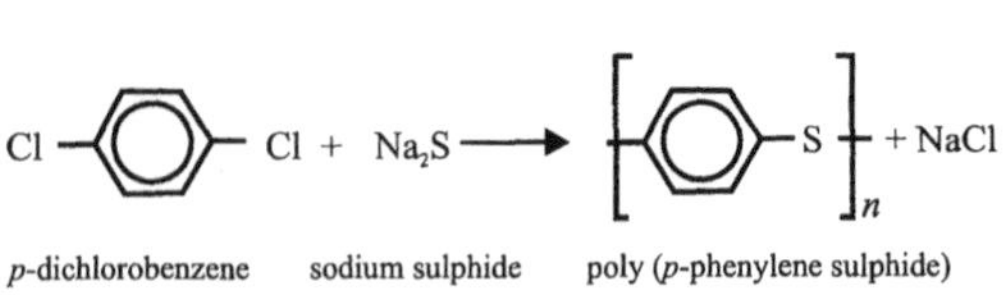

p-dichlorobenzene   sodium sulphide   poly (*p*-phenylene sulphide)

Fig. 4.25 Synthesis of poly (*p*-phenylene sulphide).

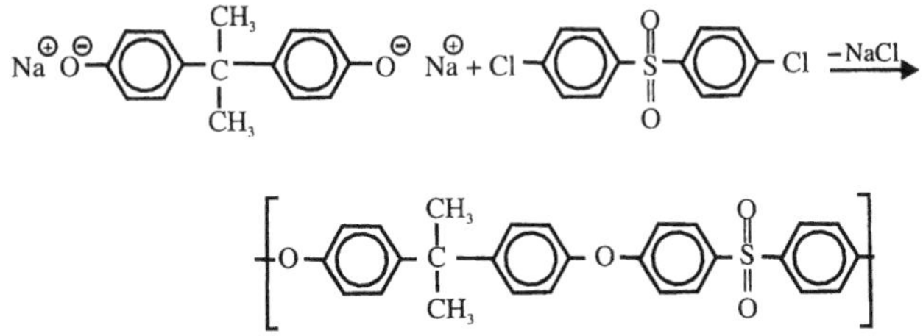

Fig. 4.26 Typical route to aromatic polyether by nucleophilic substitution (shown here for a polyether sulphone).

500 °C and thus is not processable; thus other functionalities are incorporated into the chain to make it less stiff. Usually this is an ether linkage and commercial examples include Udel™ and Victrex® variously prepared from halodisplacement as shown in Fig. 4.26.

Starting materials are not cheap, but the polymers are good stable engineering thermoplastics. Importantly they are permeable and can be used in membranes and separators. They are notably biocompatible and stable to sterilization conditions. They are also resistant to radiation (as are a number of aromatic polymers) and these applications are of sufficiently high added value to influence the economics of manufacture.

Ketones can also be employed instead of sulphones. Here the polymers (e.g. PEEK) are considered to be 'High Performance' ones and are treated in Chapter 6.

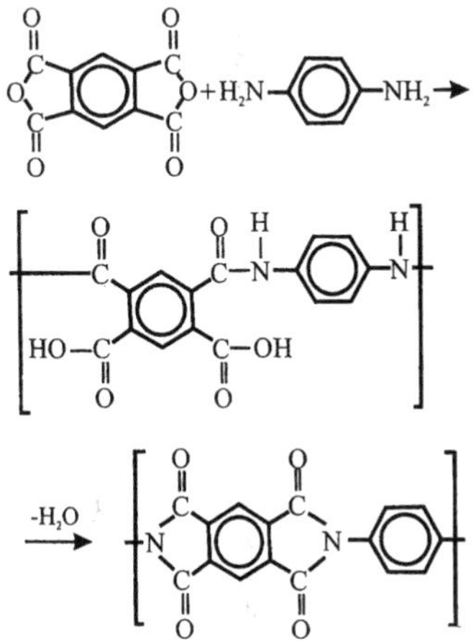

Polyimides and related polymers
Good temperature performance has been noted for a number of step-growth polymers, but the best ones are polyimides and other fused-ring or ladder polymer systems. The specialized application makes these economic despite some chemical sophistication. Polyimides are made in two steps of differing reactivities (Fig. 4.27). There is not space in this *Primer* for a full discussion and thus indicative equations are given.

A commercial polyimide may also have ether units in the chain to help processability. This is typical of an industrial polymer which combines benefits from two different linking units. (Note again the use of bisphenol A) (Figure 4.28).

Figure 4.29 shows a polybenzimidazole, while a polyimidazopyrrolane and a polybenzobisthiazole are shown in Fig. 4.30. The raw materials are not cheap but the polymers are cost-effective for specialized applications. A key

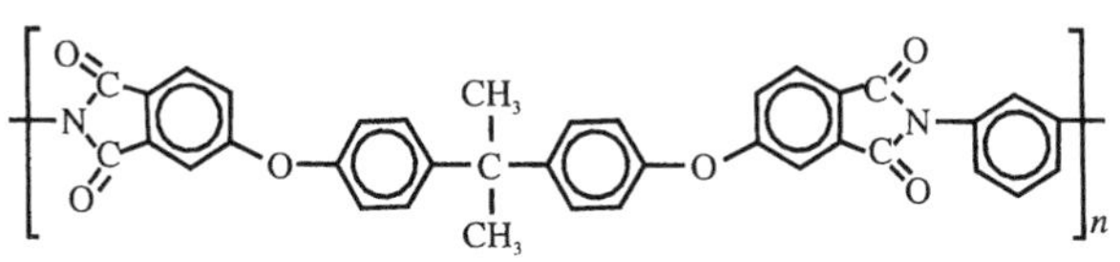

Fig. 4.27 Formation of a polyimide (note first attack on an anhydride followed by second attack on a carboxylate, with loss of water in second step only).

Fig. 4.28 Commercial polyetherimide.

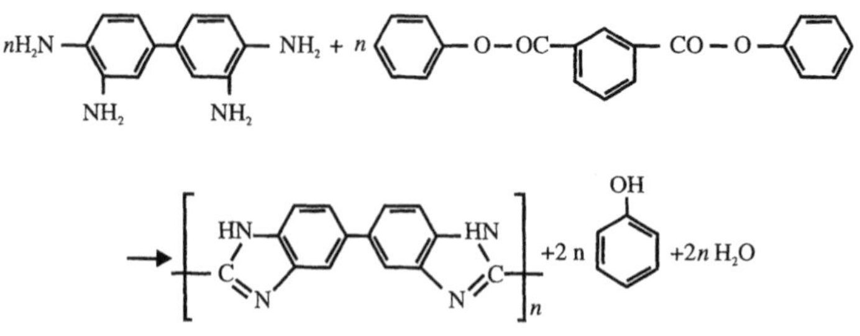

Fig. 4.29   Commercial synthesis of polybenzimidazole.

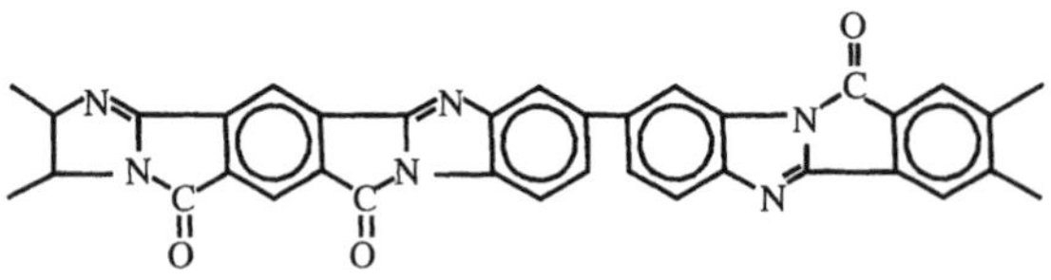

(a) polyimidazopyrrolane

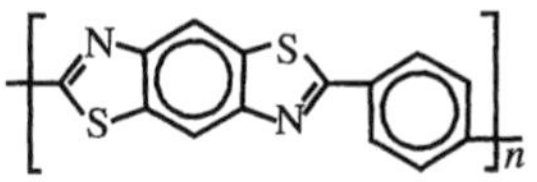

(b) poly(p phenylene benzobisthiazole)

Fig. 4.30   Temperature-resistant fused ring polymers.

benefit is that they can withstand temperatures even greater than 500 °C. This is excellent performance from organic-based materials and shows how far modern polymers have extended the range of desirable properties from the earliest simple systems.

5   Three-dimensional networks

5.1   Introduction

Three-dimensional (3-D) networks are the toughest and most rigid materials, since the polymer chains are linked together in all directions to give effectively a single giant molecule. Stone is a natural inorganic ionic-linked 3-D network, whereas polymer 3-D networks are generally held together by covalent bonds. The bridging units are called cross-links and, while in the previous two chapters it has been convenient to divide linear polymers mechanistically into chain- and step-growth types, networks can be hybrids, employing either or both systems.

Only relatively few cross-links per chain are needed to provide a useful material, since too many cross-links produces a brittle powder under internal stress. If the material requires a particular shape for an application then this must be done before the final cross-linking takes place because, once 'set' into its final form, the cross-linked polymer can only be sawn or cut into shape. Three-dimensional networks do not melt, although segments may go through phase changes with temperature. They are insoluble, although lightly cross-linked ones can be solvent-swollen. They are therefore prepared in two stages, the first giving a processable intermediate that becomes the intractable final product in the second one. These principles are demonstrated for several systems, including Bakelite (phenol–formaldehyde polymers) (which is fully step-growth) and fibre glass (linear unsaturated polyesters) (which is a step-growth backbone with chain polymer cross-links), and also the vulcanization of rubber (which is a chain polymer backbone with chain cross-links). Polymers which set hard after heating, usually because of a thermal cross-linking reaction, are called thermosets.

5.2   Formaldehyde (methanal) systems

This is the oldest fully artificial polymer system, produced by Leo Baekeland in 1907 and known as Bakelite, although the reaction of phenol with aldehydes had been studied earlier. In a successful marketing ploy Bakelite was patented as 'fireproof celluloid' to capitalize on the dangerous flammability of the existing material.

Two procedures may be followed, which lead to an intermediate soluble and fusible (meltable) resin by attack of a phenol-derived species on the carbonyl-derived species. Baekeland's original base-catalyzed technique is called the Resol process, while the newer acid-catalyzed Novolak process is widely used. Both processable intermediate resins are converted to the final cross-linked solid by acid-catalyzed second stages.

The chemistry of **Resol resins** involves attack of the phenolate anion upon the polar carbonyl group in the aldehyde as shown in Fig. 5.1. The next steps

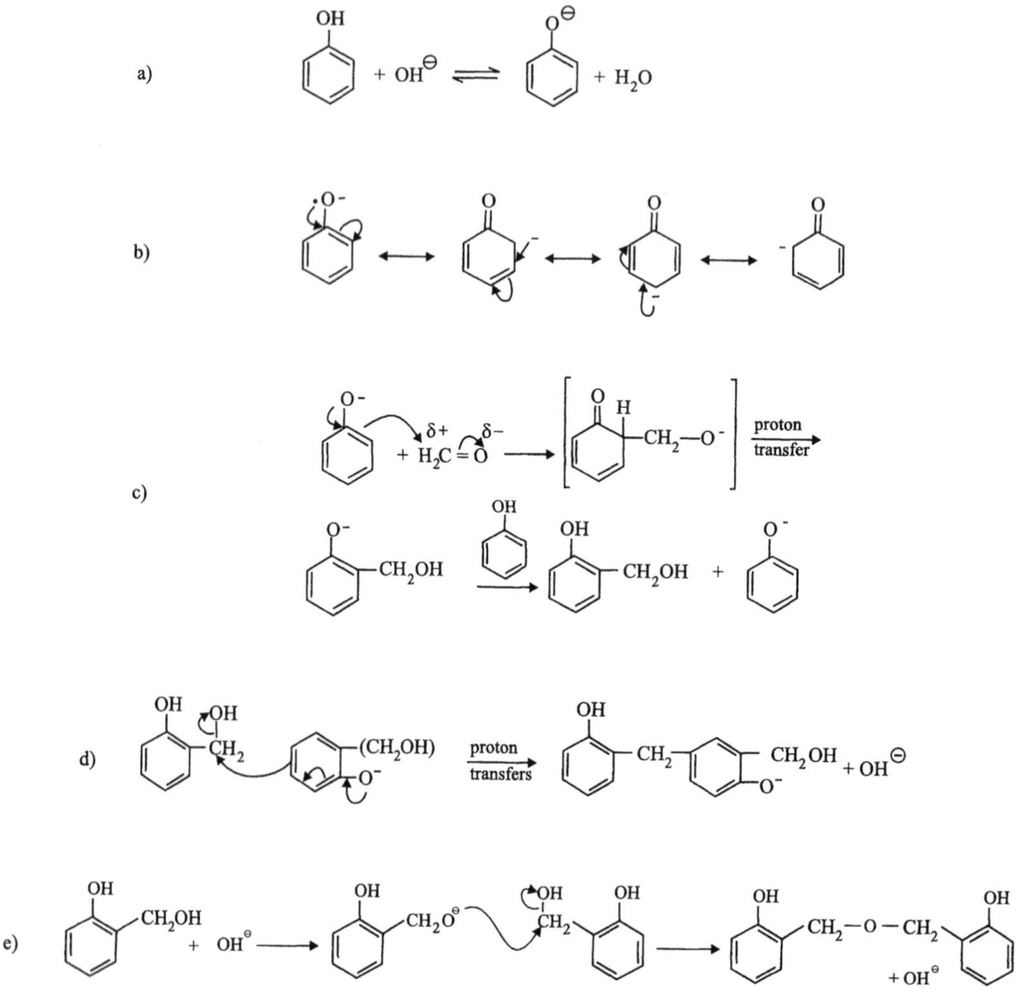

Fig. 5.1   Base-catalysed reactions of methanal with phenol (a) formation of phenolate anion; (b) resonance forms of anion showing reactive ring positions; (c) primary ring-substitution reaction; (d) formation of methylene bridges; (e) formation of ether bridges.

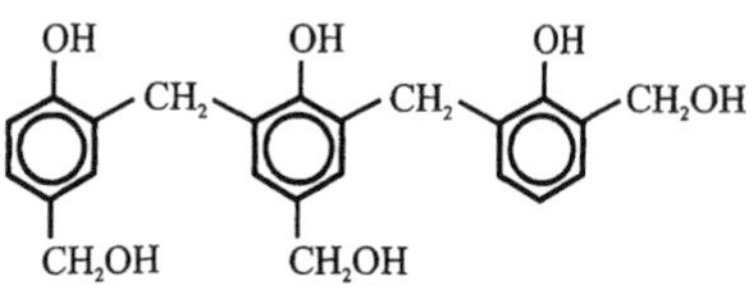

Fig. 5.2   Possible structure of resol resins (from base-catalysed reaction of excess methanal with phenol) (methylene bridges shown only).

involve nucleophilic displacement of hydroxy from the hydroxymethyl group, which can lead to either methylene bridges between rings or ether linkages between rings.

The opportunities for cross-linking occur because phenol can react at ortho and para positions and the whole point of the two-stage process is to limit the first stage of reaction to avoid insoluble gel formation. There is an excess of methanal in the mixture so the end-groups in the intermediate resin are CH_2OH, and conditions are controlled to avoid over-reaction. A sketch of possible resin structure is shown in Fig. 5.2. To form the final 3-D network in a separate second step the processable resin is made slightly acid and heated to 150 °C.

All insoluble and infusible materials are hard to analyse since solution phase analysis (NMR, liquid chromatography) or vapour phase techniques

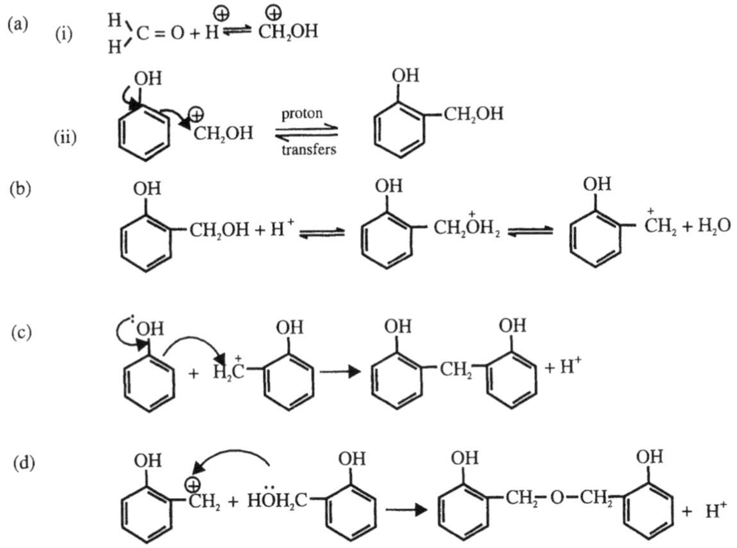

(a) (i) $\overset{H}{\underset{H}{\diagdown}}C=O + \overset{\oplus}{H} \rightleftharpoons \overset{\oplus}{CH_2OH}$

(ii)

(b)

(c)

(d)

Fig. 5.3 Acid-catalysed processes in phenol–methanal systems (a) (i) protonation of methanal (also via hydrate in aqueous medium); (ii) primary ring substitution reaction; (b) dehydration of methylol species; (c) formation of methylene bridges; (d) formation of ether linkages.

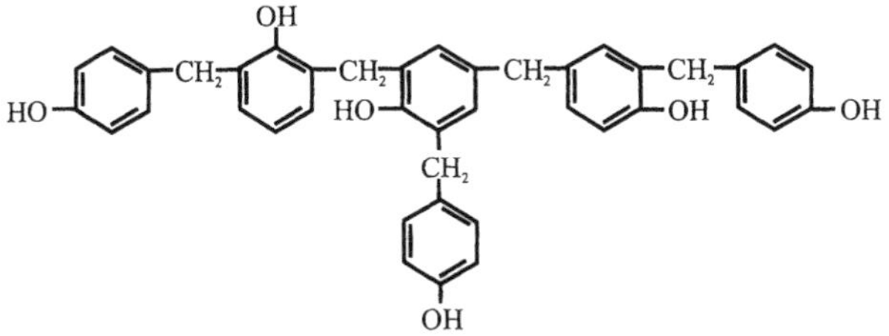

Fig. 5.4 Representation of Novolak phenol–methanal resin (formed by acid catalysis with excess of phenol).

(mass spectroscopy, etc.) are ineffective. Thus chemical reactions in these 3-D networks are not known with certainty. The acid-catalyzed processes that may occur in phenol–methanal systems are shown in Fig. 5.3 for the **Novolak** process, where excess phenol is used in the starting mixture.

This gives an intermediate resin of possible structure shown in Fig. 5.4. With all methanal consumed the end-groups cannot react further until a cross-linking agent is added, which is a benefit of the Novalak system since different species such as hexamethylene tetramine, itself made separately from methanal and ammonia (Fig. 5.5), can be used instead of simply adding more methanal. This can be used to affect final properties.

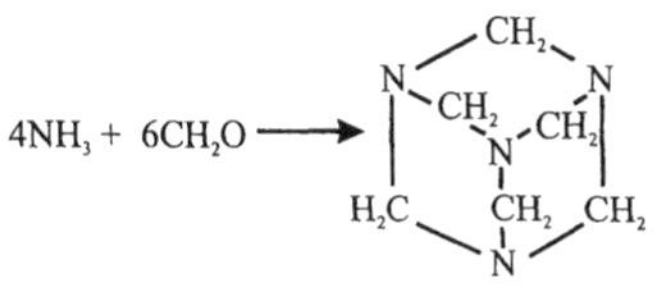

Phenol–formaldehyde resins are chemically stable to all but strong oxidizing acids, although they are more easily attacked by bases. They have good electrical insulation properties and were once ubiquitous as the brown-coloured electrical plugs and adaptors since they are easily processed into complex shapes. They also form laminates and composites with many materials and

$4NH_3 + 6CH_2O \longrightarrow$

Fig. 5.5 Synthesis of hexamethylene tetramine.

fillers, particularly cellulose-based biorenewable materials such as wood, cotton, fabric or paper. These composites are tough and have tensile strengths in excess of 40 MPa.

Other methanal–amine resins

The basis of the phenol–methanal chemistry described above is acid- or base-catalyzed attack of a nucleophile upon the carbonyl group of methanal. Phenol can be replaced as a nucleophile by a reactive amide or polyfunctional amine, which produces a similar network system. These are two main examples: urea–methanal and melamine polymers, although other systems such as urea–casein have also been used commercially.

Urea–methanal

It was known in the 1880s that urea reacts with methanal (formaldehyde) to give resins, but moulding systems required the development of acidic accelerators in the 1930s. The present material is used in electrical fittings, caps for plastic bottles and other rigid mouldings, as well as adhesives and coatings (in versions with lower RMM), while cellulose compatibility has resulted in large scale use onwards from the 1950s in reconstituted wood products, particularly chipboard. Another use, in a foamed form as 'cavity wall insulation', started in the 1960s, but is decreasing because of concerns over slow release of methanal from the *in situ* prepared foams. Factory-prepared materials are less affected, since devolatization (accelerated removal of volatile species) occurs during the high-temperature processing of most polymers (e.g. during extrusion, moulding, etc.).

Urea (made from ammonia and carbon dioxide at 200 °C and 300 atmospheres) is heated with methanal (as aqueous formalin) in first mildly alkaline then acidic conditions. The two-step reaction system goes via a low-RMM resin (itself a useful adhesive), before finally the 3-D network is formed as represented in Fig. 5.6.

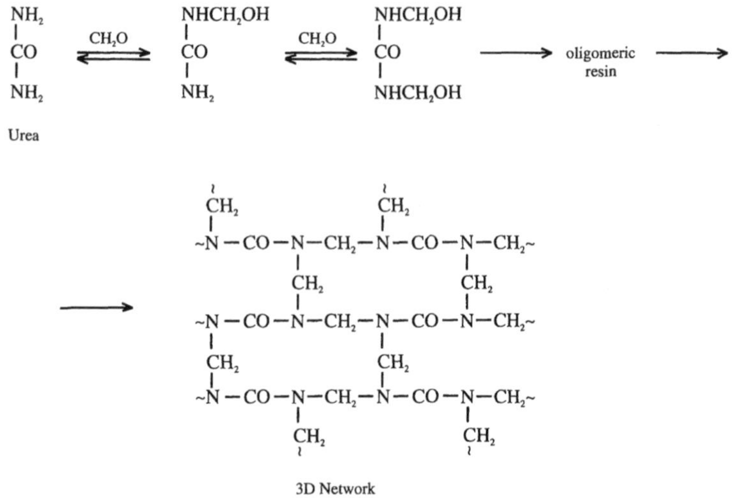

Fig. 5.6 Representative formation of a urea–methanal network polymer.

A useful example of industrial fine-tuning of a process is that resin stability and miscibility can be altered by end-capping with an alcohol, as shown in Fig. 5.7 for butanol capping.

Urea–methanal resins are tough (> 50 MPa tensile strength when filled with cellulose or glass) and, unlike the phenol analogues, may be produced in a range of colours. A useful feature of many polymer applications is that uniform coloration may be maintained throughout the bulk of an article. So whereas wood, for example, is painted on the surface and rapidly shows up scratches and abrasions, a plastic chair does not do so.

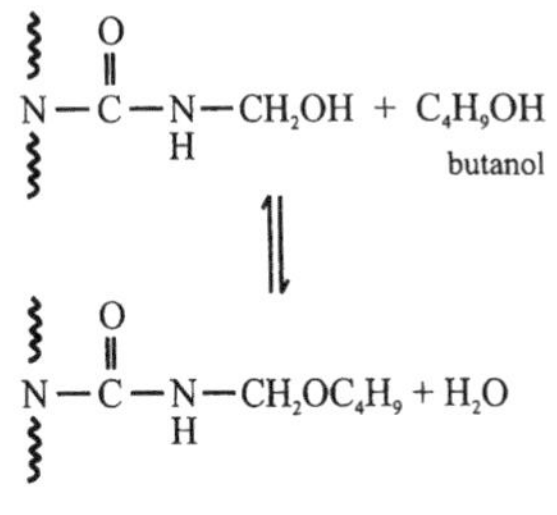

Fig. 5.7 Butanol capping in urea–methanal networks.

Melamine

This methanal-based system has a surface energy such that it is stain-resistant, particularly to tea, coffee and foodstuffs, and it thus became popular in the 1950s as a coating material in the catering industry. It is also easy to colour, with good thermal and chemical stability and is used as domestic mouldings in tableware, handles for irons and other decorative consumer products.

Melamine is the trivial name for a cyanuric derivative. This class of compound contains a six-membered carbon–nitrogen ring that has some similarities to benzene. Other derivatives are widely employed, e.g. triallylcyanurate (TAC) (cf. Fig. 5.14). These compounds may look complicated, but they are readily formed by trimerization of species obtained from starting materials that are already available to polymer companies. Thus for melamine there is dehydration of urea and trimerization of the resultant cyanamide, as shown in Fig. 5.8.

If melamine is considered as a sophisticated variant of $R-NH_2$ then chemistry with methanal via methyl derivatives, onwards through methylene bridges and ether bridges can easily be envisaged. The practical process is in two steps, first heating to 80 °C in mild alkali (aqueous carbonate) with an excess of methanal over melamine to give the intermediate resin as a syrup, followed by heating at a higher temperature to give the solid network. Butanol can also be employed to alter properties, particularly for surface coatings, in a similar manner to urea–methanal systems.

Finally, a further variation on methanal-based networks was described in the 1890s and became popular in the 1930s (as 'Erinoid') for buttons, door handles and other decorative coloured articles with a shiny finish. This was manufactured by reaction of methanal with casein, which is a protein found in milk. This demonstrates a useful hybrid between an artificial component and a

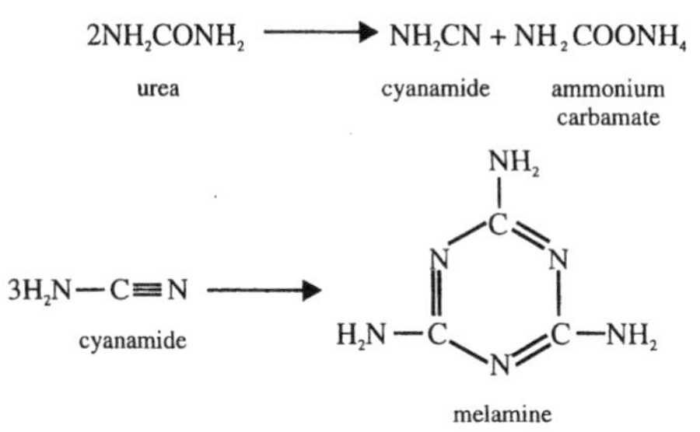

Fig. 5.8 Formation of melamine from urea.

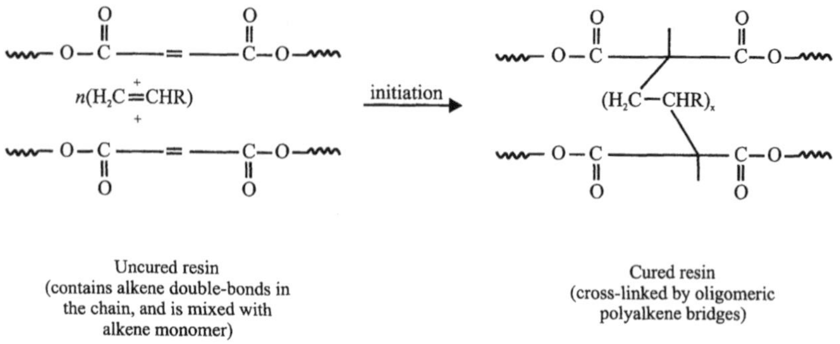

Fig. 5.9 Schematic of the principle for cross-linking linear unsaturated polyesters ('fibre glass').

readily available natural material. In the early 1900s polymers became popular for decorative items, replacing tortoiseshell, ivory, horn, and stones, for example as 'artificial jade'. The control of shape by processing allowed many novel items to be made.

5.3 Linear unsaturated polyesters ('fibreglass')

A processable polyester chain with a relatively low RMM (a step-growth resin) with double bonds deliberately placed in the backbone can be subsequently cross-linked to a 3-D network by chain polymer chemistry, as shown schematically in Fig. 5.9.

This type of polyester, and the great strength enhancement due to glass-fibre reinforcement, resulted in Fibreglass becoming commercially available on a wide scale in the 1950s leading to usage as large-scale mouldings for boat hulls, lorry (truck) cabs, car bodies, roofing panels and other structural components, as well as the familiar fibreglass bodywork repair kits for cars.

The polyester backbone needs components that are internally compatible and also compatible with the alkene derivative later used for cross-linking. Unsaturated diols are more expensive than unsaturated acid derivatives, so the normal system is to use propylene glycol (propane, 1,2 diol) in which the methyl group produces greater organic compatibility and is also necessary to limit crystallinity, since the highly ordered crystallinate regions would hinder thorough microscopic mixing between components of the two-part cure system. The unsaturation is in the acid derivative. Maleic acid is readily available. The anhydride is preferred because this reacts to give the two ester linkages with the expulsion of only one molecule of water (which must be removed to drive over the equilibrium). In addition, the anhydride is more reactive than the acid, to help the first step, although the second step is normal esterification of an acid group as shown in Fig. 5.10. (This is a general feature of nucleophilic attack upon anhydrides, and also applies in reactions leading to polyamides and other polymers. See, for example, Fig. 4.27.)

A saturated acid derivative is present to limit the available double bond sites to avoid over cross-linking. For economic reasons and for compatibility in the mixture phthalic anhydride is often used.

Both the anhydrides are made by variants of a catalyzed industrial reaction that would not be employed in the small-scale science laboratory. For maleic

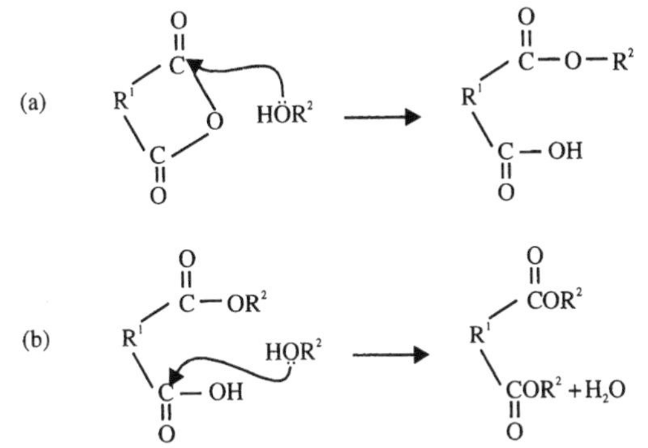

Fig. 5.10 Formation of diester by attack of an alcohol upon an anhydride (a) monoester formation, no water molecule expelled; (b) diester formation, water molecule evolved.

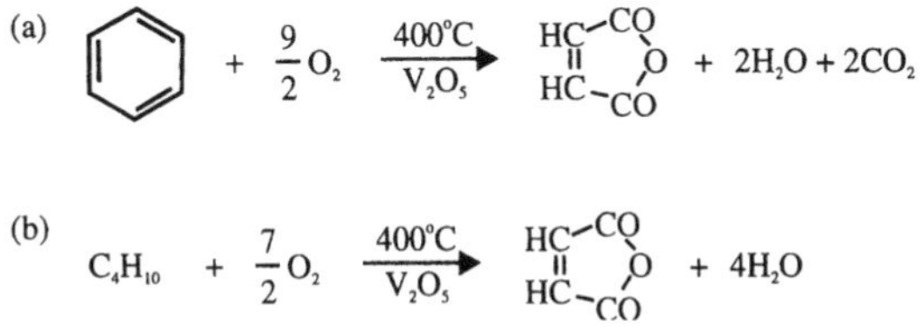

Fig. 5.11 Commercial synthesis of maleic anhydride from (a) benzene; (b) butane.

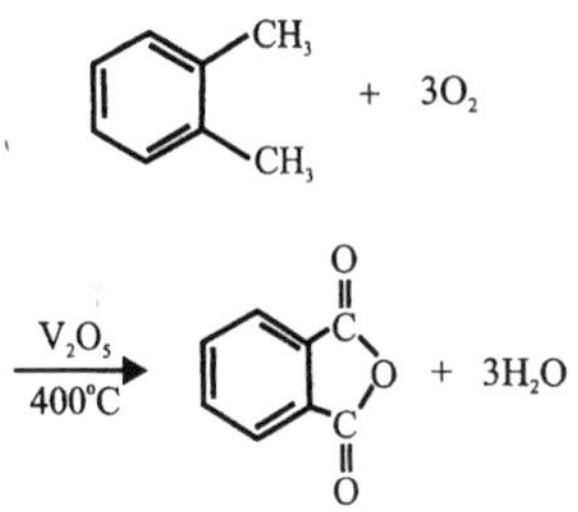

Fig. 5.12 Commercial synthesis of phthalic anhydride from *o*-xylene.

anhydride, benzene (cheap from petroleum) or more recently *n*-butane is directly oxidized with air (cheap) in the vapour phase at ~400 °C over a vanadium pentoxide catalyst (Fig. 5.11). For phthalic anhydride aerial oxidation of xylene (previously naphthalene) is used (Fig. 5.12).

Resin preparation
The components are reacted together to give an oligomeric resinous polyester whose structure is represented in Fig. 5.13.

After resin formation the cross-linking agent is then added for the second stage. Styrene can be used when the fibre glass is prepared and cured at the factory, but for applications where there are environmental concerns, a less volatile alkene such as diallylphthalate or TAC is used (Fig. 5.14).

The cross-linking reaction is set off by a free radical initiator, such as benzoyl peroxide, which thermally decomposes to give aryloxy and then aryl radicals. There are also halogenated variants, or alkyl variants such as lauroyl peroxide (perdodecanoic acid). Cheaper but less well defined peroxide mixtures such as those obtained by aerial oxidation of methyl ethyl ketone or

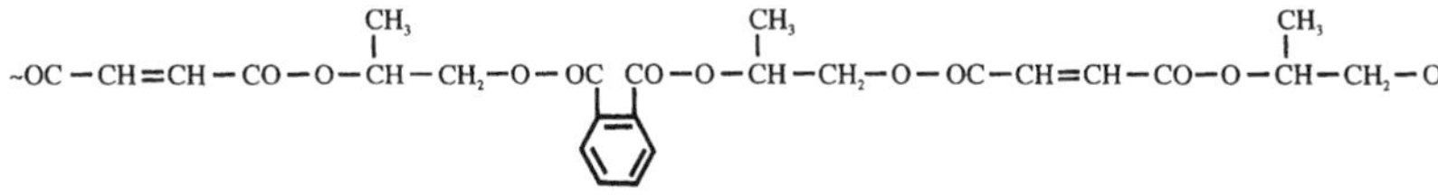

Fig. 5.13 Representative structure for segments of a linear unsaturated polyester chain. The double bonds in the product are a mixture of *cis* and *trans* isomers.

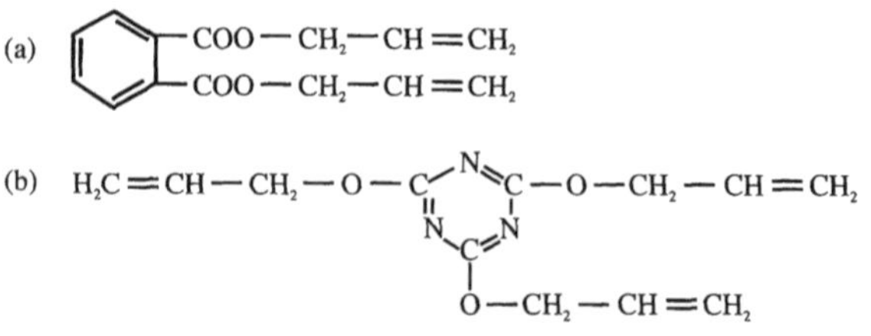

Fig. 5.14 Cross-linking agents for linear unsaturated polyesters (a) diallylphthalate; (b) triallylcyanurate.

Table 5.1 Typical values for various properties of cured linear unsaturated polyester, unfilled versus reinforced

	Unfilled polymer	Glass filled
Tensile strength (MPa)	60	800
Flexural strength (MPa)	120	1000
Impact strength (J/m)	100	3700
Glass content (wt. %)	0	70

cyclohexanone (see Chapter 3, for example Fig. 3.3) are often employed in industry.

Accelerators are added to control peroxide breakdown. These are usually transition metal salts, usually of organic anions such as octoates or naphthenates for organic compatibility in the polymer matrix (see Chapter 3). The colour of, for example, cobalt ions can be seen in two-part automobile bodywork repair kits. The ability to cross-link at room temperature is useful for these small-scale kits, while accelerators are also useful to avoid problems from differential heating from factory-prepared large items such as boat hulls.

A key aspect of fibre glass is the strength reinforcing effect of glass fibre, which is mixed in before final cure of the resin. Exact properties vary with the nature of the filler, as shown in Table 5.1, and needle-like particles have a better effect than spherical ones. Glass has hydroxy groups on its surface, and polyester systems seem particularly compatible (all fillers have compatibility requirements with the matrix polymer and choice of filler for any purpose can be a subject of some deliberation). Note the tremendous increase in strength in the glass-loaded systems. All entries in Table 5.1 concern the same composition of matrix polymer.

Cured fibre glass is very strong, is mostly unaffected by reagents apart from strong alkalis and oxidizing agents, is moderately temperature-resistant, the good physical properties degrading above 100 °C (except in a temperature-resistant grade).

5.4 Alkyd resins (polyester paints)

This class of cross-linked coatings remains widely used and is based on glycerol as the alcohol component of a polyester. Glycerol is a triol and straightforward reaction with phthalic anhydride directly produces a cross-linked network. 'Glyptal' resins, produced by General Electric in 1912, have a

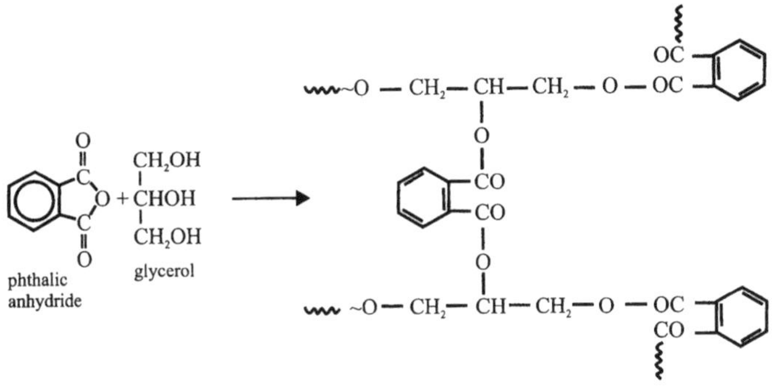

Fig. 5.15 Formation of a glyptal cross-linked polyester resin.

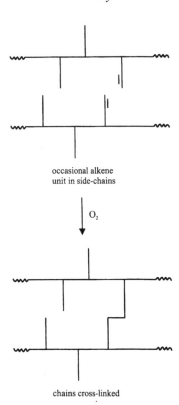

occasional alkene
unit in side-chains

chains cross-linked

Fig. 5.16 Schematic for cure of an alkyd resin paint upon exposure to oxygen.

typical structure as shown in Fig. 5.15. An early use for this was bonding mica into sheets for electrical insulation, but in general these coatings were prone to over cross-linking since the system forms its cross-links during the chain-building esterification reaction.

A more sophisticated variant employs a two-stage process in which esterification gives a material which subsequently cross-links in a chain-growth reaction. This allows the processable resin intermediate to be manipulated before coming to a final intractable network, in a process analogous to that employed in fibre glass, but in this case the polyester chain is saturated and the alkene double bonds are in the pendant side chains. No extra cross-linking agent needs to be added, and 'alkyd' resins were first produced in the 1930s to be widely used as coatings and particularly as paints. A schematic of the principle is given in Fig. 5.16.

Glycerol is found naturally as a triester containing some unsaturated acid groups and the triester is mixed with phthalic anhydride and free unsubstituted glycerol triol in correct proportions. Transesterification (see Fig. 4.2) 'scrambles' the acid-derived side chains to give a resin with some saturated and some unsaturated side chains attached to each triol unit. The resin is dispersed in a solvent and kept hermetically sealed in a tin but, once painted onto a surface and the solvent allowed to evaporate, oxygen from the air can attack the alkene groups and cause cross-linking. Only a low cross-link density is needed for the paint to set. It is now irreversibly altered and will not redissolve in the original solvent.

Hydrogen abstraction, disproportionation and other free-radical and auto-oxidation processes occur so that even saturated pendant chains can become involved in the cure, and catalysts and accelerators may be added for enhancement.

5.5 Vulcanization

This was arguably the most important early development in polymer technology, demonstrating the principle of enhancement of properties of a natural material (see Chapter 1). Originally the term referred to the heating of natural rubber with sulphur to give a tough elastomeric material which

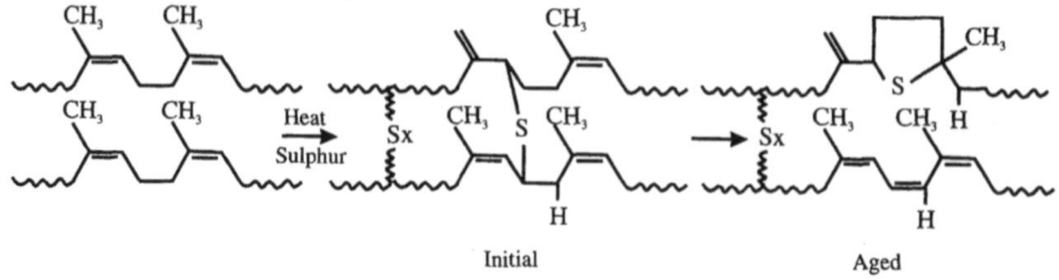

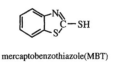

Fig. 5.17 Possible structures during initial vulcanization of rubber and later ageing processes.

eventually provided the technological breakthrough for the pneumatic tyre, and hence the growth of the automobile industry with its great effect on all aspects of modern life. However, the term 'vulcanization' is also used to describe other means of cross-linking diene-based chain-growth polymers and has even spread to the curing of step-growth silicone polymers.

Natural rubber (cis-poly isoprene)

Sulphur remains the most convenient and cheap reagent, although initiators such as peroxides are now also used. Figure 5.17 gives a flavour of the complex processes that can occur, with polysulphur bridges of differing lengths, the possibility of double-bond migrations and involvement of the methyl groups of the polyisoprene (poly 2-methyl butadiene). Ageing effects in the final product also occur with formation of cyclic structures in the chain and other secondary processes.

As always with modern commercial processes there has been significant development and the 'accelerated sulphur vulcanization' system now used is significantly more complex than simply heating rubber with sulphur to 150 °C, where reaction is relatively slow. The 'accelerators' are sulphur-containing compounds such as mercaptobenzothiazole (MBT) or other molecules such as diphenylguanidine (DPG) (Fig. 5.18), together with an 'activator' which is a mixture of zinc oxide and a surfactant for compatibility. Other additives such as carbon black (for car tyres) may be necessary for the application. All components are blended together and heated. Details of the chemistry are not all understood, but this does not detract from the continuing practical use of this technologically important process.

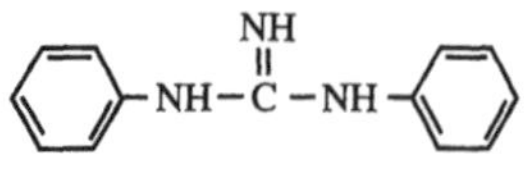

As in all networks the degree of cross-linking affects pliability and physical properties, and highly cross-linked vulcanised rubber gives the tough rigid material 'ebonite' ('vulcanite').

mercaptobenzothiazole(MBT)

diphenylguanidine(DPG)

Fig. 5.18 Typical accelerators for rubber vulcanization.

Chloroprene (poly 2-chloro 1,3-butadiene) and chlorosulphonated polyethene

Chloroprene is an artificial rubber where the methyl group is replaced by a chlorine atom (see Chapter 3). The electronegative chlorine interferes so that the reaction with sulphur is ineffective. Instead the term vulcanization is used for cross-linking of this material by heating with zinc and magnesium oxides. The chemistry is complex but rearrangements and chlorine removal lead to ether cross-links. A similar strategy works for other halogenated polymers such as chlorosulphonated polyethene elastomers.

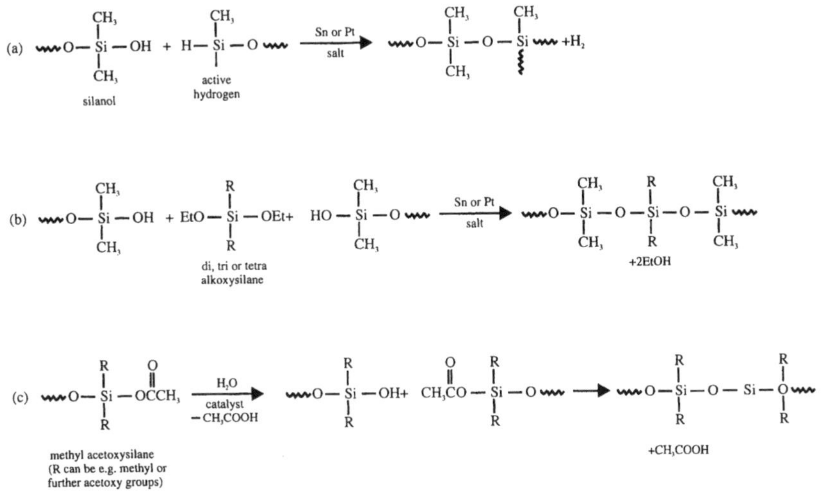

Fig. 5.19 Representative reactions for vulcanization of silicones (a) using reactive Si-H bonds in copolymers; (b) using alkoxysilane cross-linking agents; (c) using acetoxysilanes (one-part system).

Silicones

These are step-growth polymers but the term vulcanization is applied to their cross-linking reactions, which are shown in Fig. 5.19. These are often two-part systems and are effective at room temperature. Thus if a siloxane copolymer with a reactive Si–H bond is mixed with a hydroxy-ended analogue, hydrogen is evolved and the chains are joined. Alternatively, two silanol-ended chains are reacted with an alkoxysilane to join the chains (shown for dialkoxy, but tri- and tetra-alkoxy give more opportunities for cross-linking), while a useful one-part silicone vulcanization system involves protecting the silanol as an acetoxysilane ester. This is stable until exposed to moisture whereupon Si–OH groups are liberated to join chains by directly attacking other acetoxysilane groups.

5.6 Other chemically cured network systems

There is a wealth of chemistry available for the formation of cross-links in polymer systems. It is worth mentioning a number of systems widely used as tough coatings; these include thermosetting acrylics and epoxies (also used as adhesives), as well as photocured systems.

Thermosetting acrylics/epoxies

The oxirane ring (epoxide) is readily opened by a nucleophile. This is the basis of epoxyresin adhesives and is shown schematically in Fig. 5.20. Adhesives involve the joining together of resinous components, to give final structures with high RMM, and reaction may occur at end-groups or at trifunctional junction points in mid-chain.

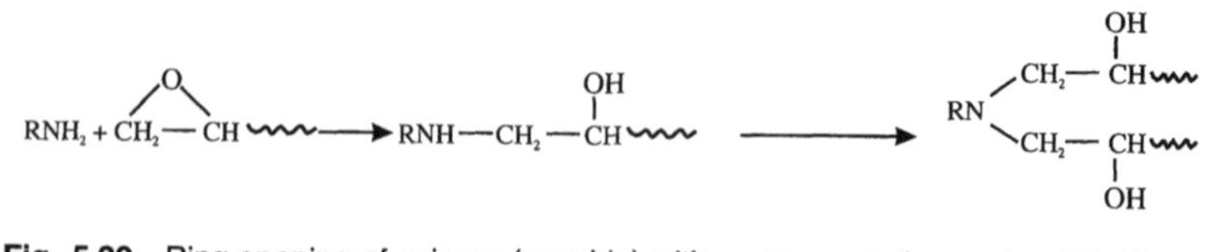

Fig. 5.20 Ring opening of oxirane (epoxide) with a representative nucleophile (here RNH$_2$) which can react twice.

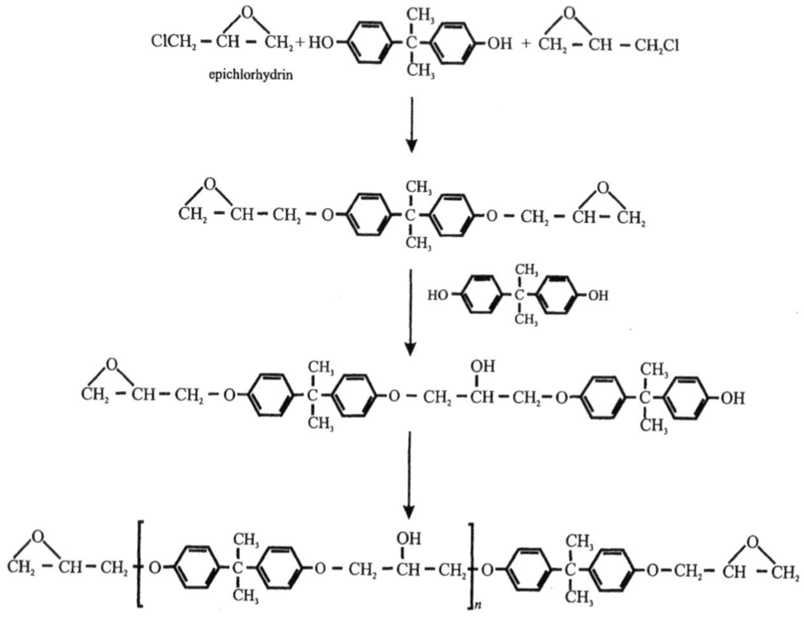

Fig. 5.21 Formation of a bisphenol A epoxy resin.

Epoxy resins

The epoxy species can be added to a step-growth polymer by reaction of epichlorohydrin with bisphenol A as shown in Fig. 5.21. This gives epoxy-ended polyether chains with hydroxyalkyl internal units from ring opening of the oxirane. The epoxyether end-groups are called 'glycidyl' units in the industry. Epoxy resins were produced by CIBA–Geigy in 1943 and are usually cured by a polyfunctional amine or an anhydride as nucleophile. Again the practical system involves promoters, accelerators and other additives.

Thermosetting acrylics

Here an all-chain polymer system is used, often a tri-block copolymer containing a strength-conferring section (acrylonitrile, styrene, methyl methacrylate), a more flexible unit (ethyl or butyl acrylate esters), and a unit with necessary functionality (acrylic acid for COOH, hydroxyethyl acrylate [$-CH_2CH_2OH$] for pendant hydroxy groups, or glycidyl acrylate which contains the oxirane ring). Depending on the structure of the tri-block these can either self-react upon heating or else additional cross-linking agents (epoxy-ended resins for nucleophilic chains, or polyfunctional amines appropriately) are used to facilitate the cure. Tough resilient coatings are obtained.

Photocured acrylics
These have biomedical applications, for example as the white plastic fillings for teeth used in dentistry. Photochemical reaction of activated alkenes was shown as long ago as the 1830s (see Chapter 1) and has the benefit that initiation is detached from the thermal conditions of the system. Development effort has produced acrylic/urethane composite resins for better properties and biocompatibility and dangerous ultraviolet radiation has been replaced by blue visible light.

5.7 Electron beam cross-linking

A high-energy electron beam induces reactivity upon striking a polymer in air. In particular free radical-type processes can occur, leading to cross-linking of a saturated chain without need for deliberate introduction of chemically reactive sites.

This was exploited commercially in the 1950s by Raychem, who targeted electrical wiring for aircraft as a specific market. Previously plastic insulation had been relatively heavy and unwieldy, and electron beam cross-linking shrank the plastic down so that better insulation was obtained from lighter and tougher coatings, an immediate benefit given the extensive amounts of wiring in an aeroplane. Passing wire at speed through the beam allowed kilometre-scale lengths to be easily processed.

A further development by this company was 'heat-shrink' plastics. It was found that the low levels of cross-linking desirable for most applications of polymer networks do not greatly affect phase changes in the sections between cross-links (though the concept of melting point is not directly applicable). If, therefore, a moulded object is exposed to an electron beam, and is then heated above its glass transition temperature (T_g), stretched and rapidly quenched by cooling, it remains 'frozen' in an expanded form. Subsequent heating above T_g will cause the article to snap back into the original shape it had when it was exposed to the beam. This allows, for example, telephone splice cases to be fitted in the expanded form, then heated to shrink to the exact dimensions around a wiring junction. Other connectors can be protected in a similar way.

A number of chemical processes can occur in the electron beam and a wide range of polymers, both chain-and step-growth, will react, the exceptions being polymers containing aromatic rings or other extended conjugation systems, where beam energy becomes dissipated with loss of effect.

Cross-linking by X-rays and γ radiation is also feasible, though it is expensive and subject to regulatory control because of the use of radioactivity.

5.8 Physical cross-linking

Network systems tend to be thermosets, in which a cure produces new covalent chemical bonds to link chains permanently, but there are also physical methods of cross-linking that are reversible. To some extent all thermoplastics (tough when cold, fluid when hot) behave as a result of physical interactions between their long chains, but there are some more specific systems. For example, ionic bonds between pendant acid groups on a polymer (e.g. polyacrylic acid) and a polyvalent metal cation will hold chains together, particularly in a low dielectric environment that promotes ion-pairing.

Polar forces are also important and edible jelly is held together by sufficiently strong hydrogen bonds in a gelatin matrix such that water in quantities many times the mass of the matrix can be trapped in the gel which still feels relatively dry to the touch.

A neat example of physical cross-linking is found in block copolymers that have one section amorphous and another section crystalline, such as styrene–ethene copolymers. In this case, in the solid the crystalline regions contain many chains fitted into regular arrays and held tight together in a microcrystallite, while the amorphous regions are disordered and allow some chain motion. Such polymers can also trap solvents in the amorphous regions to become gels. However, when heated above the melting point of the crystal region the polymer immediately flows giving a reversible cure on cooling down again.

6 Functional polymers

6.1 Introduction

Commonly the emphasis for polymer applications has been based upon the physical and mechanical properties arising from high molar mass (RMM). The deliberate enhancement and exploitation of some other properties has been less widely studied, although the presence of chemical functionality in many polymer systems has been recognized for a long time. In Chapter 5 it was shown that the presence of reactive alkene functionalities in the polymer chain allows the vulcanization of natural rubber, a strategy later adapted to an artificial polymer system in linear unsaturated polyesters (fibre glass), with a variation employed in the drying (curing) of certain paints (alkyd resins) where oxygen in the air initiates free radical cross-linking of pendant alkene functionalities. In general, the chemistry of adhesives involves the presence of groups that can react with each other.

The use of cellulose-based polymers arises from reaction of the hydroxy groups on the sugar rings (see Fig. 1.3) and the properties of nylons can be manipulated, for example by post-reaction with aldehydes. However, the specific search for added value polymer behaviour has been a relatively recent phenomenon. A good example is the development of the water-absorbent polymers used in babies nappies (diapers). These materials are based on water-soluble polymers such as cellulose derivatives, oligosaccharides or acrylamides which are lightly cross-linked to prevent them dissolving. Instead they become swollen and can absorb more than 100 times their own weight of water. This capability originates not just from the intrinsic hydroxylic nature of the polymer chain, but also from developments in the control of porosity and permeability related to the specific application.

This chapter concerns a review of recent developments in functional polymers, selected to show the essential strategic factors in each case. Some materials have found practical uses, while others remain research curiosities at present.

6.2 High performance polymers

Principles

This refers to polymers with extreme mechanical strength and toughness. The strategy to achieve this is instructive, since enhancement of one specific property occurs at the expense of other properties and eventually the best practical system involves a trade-off between desired features and those attainable.

Many commercial polymers of moderate strength employ aromatic rings in the backbone chain, mixed with sp^3 carbon units (see Chapter 4). This combines the toughness (on the microscopic scale) of the planar, conjugated

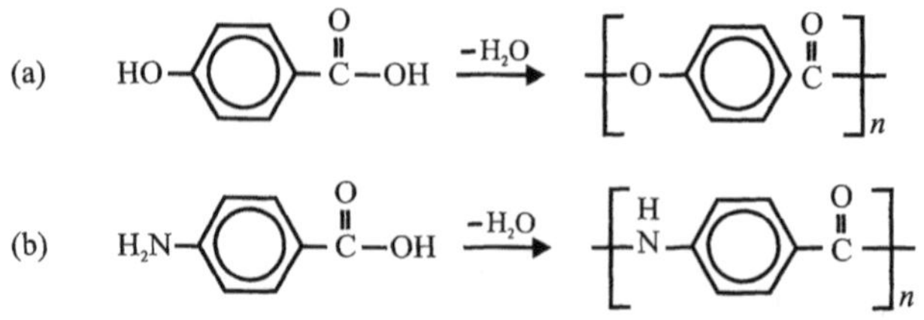

Fig. 6.1 Representative fully aromatic polymer systems (a) a polyester, poly (p-hydroxybenzoic acid); (b) a polyamide, poly (p-aminobenzoic acid).

and multiply bonded sp^2 carbon system with the flexibility of the sp^3 singly bonded carbon units (with substantial freedom around the tetrahedral carbon). A good example of this is poly(ethylene terephthalate), Terylene, while for the same reason many step-growth systems employ derivatives of bisphenol A (see page 97) which has planar aromatic rings sandwiching an sp^3 carbon unit that also possesses space-filling methyl group substituents.

The aromatic nature of the chain can be increased in, for example, poly (p-hydroxybenzoic acid) or poly (p-aminobenzoic acid) (Fig. 6.1).

These materials would be expected to be extremely strong because there is a degree of double bond character in the chain, even the single bonds, because of the resonance opportunities shown in Fig. 6.2.

The two resonances cannot be drawn with a single set of 'curly arrows'. Essentially the nitrogen lone pair conjugates in two separate directions. In reality, of course, overlap of orbitals to transfer electron density will be a simultaneous mixture of the various possible structures and thus such 'cross-conjugation' can be viewed usefully to explain the strengths of all materials of this type. Note that para substitution is needed for conjugation (ortho substitution is also suitable in principle but the chain is generally sterically hindered). A similar situation pertains in polyesters where O replaces NH in the linking unit of the chain.

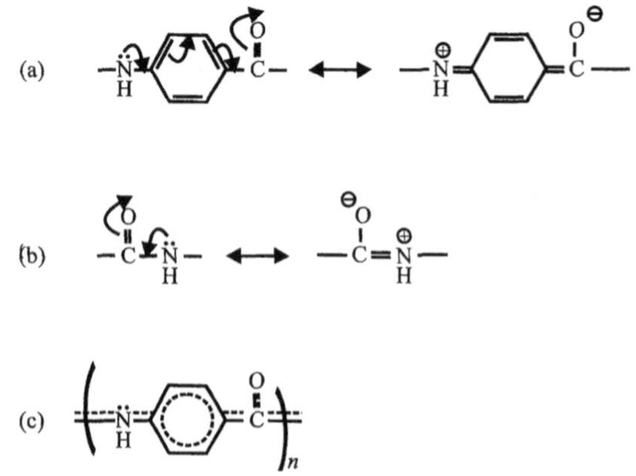

Fig. 6.2 Resonance opportunities in poly (p-aminobenzoic acid) (a) nitrogen lone-pair into ring and carbonyl group out of ring; (b) amide group internal resonance; (c) combined schematic representation.

Unfortunately, rigidity generally hinders solvation, since the co-ordination of solvent molecules to the dissolved polymer solute requires flexibility and chain motion to maximize the energy benefit from dissolution. Rigidity also hinders molecular motion in the melt. It turns out that both poly (p-hydroxybenzoic acid) and poly (p-aminobenzoic acid) are insoluble and intractable materials, with very high melting points, (in excess of 600 °C) such that they decompose before melting even under inert atmospheres. Possible benefits in toughness and strength cannot therefore be turned into reality.

However, a number of commercial high performance polymers are viable, depending on the added value of potential applications *versus* the greater production costs compared to more traditional versions of these polymers. There are several strategies, all of which involve backing off from the perfect cross-conjugated chain to trade-off strength against improved processability.

Alignment of linking units

This exploits the relationship between isomeric polymers shown by nylon 6 and nylon 6,6 (Chapter 4). Thus, while poly (aminobenzoic acid) has repeating CONH....CONH.... linkages, the equivalent chain from reaction of a diamine with a diacid can be envisaged as having the respective linkages alternating CONH....NHCO....CONH.... along the chain (see also page 6).

There are similar resonance structures to describe the cross-conjugation possibilities, and likewise for the polyester equivalent (replace NH with O in the linking unit of the chain). In both of these all bonds in the linking units are more than single, but each alternate aromatic ring is 'disconnected' from the next since two NH groups try to donate to the same ring, while two CO groups try to withdraw from the same ring.

This affects the polyamide system enough to produce the best known example of a commercial high performance polymer, Kevlar^TM, first produced by Dupont in 1972 (Fig. 6.3). (Aromatic polyamides are generally known as 'aramids' in the industry.)

Note the use of the diacyl chloride for greater reactivity to promote polymerization, which needs a solvent mixture of *N*-methyl pyrrolidone and the toxic hexamethyl phosphoramide. Kevlar^TM is not melt-processable because it decomposes below its still high melting point, but it can be processed from solution. The processing solvents are also unusual e.g. concentrated sulphuric acid. This requires use of acid-resistant apparatus which adds to the start-up costs of the commercial process.

The monomers are relatively expensive and the commercial feasibility of this polymer revolves around the high added value nature of its applications. It is best known as the polymer used for bulletproof vests, which exploits its tremendous impact-resistance. Kevlar^TM is usually solution-processed into fibres for weaving and if these are combined with an

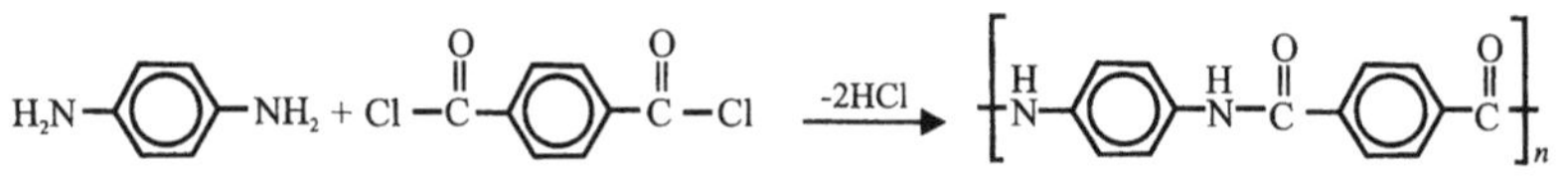

Fig. 6.3 Synthesis of Kevlar^TM high performance polyamide.

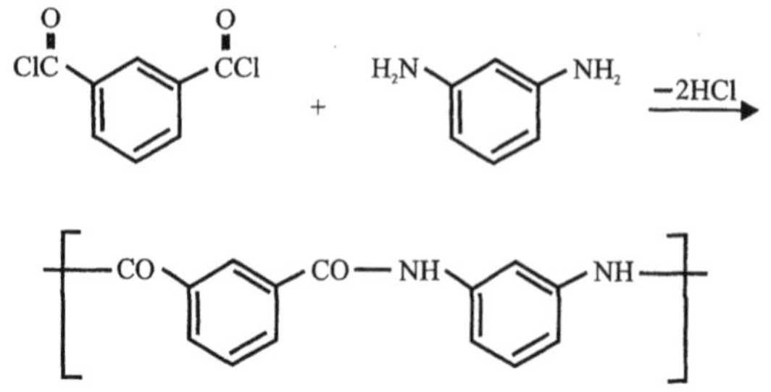

Fig. 6.4 Synthesis of Nomex™ high performance polyamide (note meta-linked aromatic rings).

appropriate conventional polymer, then more cost-effective composites can be made, although there is a trade-off between properties. Kevlar™ fibres are used to reinforce car tyres.

Aromatic ring substitution pattern and the Crankshaft Effect

The above considerations have all assumed para connections between the aromatic rings and the linking units, which maintain conjugation and tend to keep a degree of linearity in the chain, which affects crystallinity and toughness. However, when simple para-linked systems were found to be too insoluble and to have too high a melting point for practical processing, the strategy was to employ meta-linked units. This disrupts conjugation in the aromatic rings and also 'kinks' the polymer chain to affect physical properties.

The commercial polymer is Nomex™, which is older than Kevlar™, being first produced in the late 1960s by Du Pont (Fig. 6.4).

This simultaneously exploits the crankshaft effect as well as the alternate alignment of linking units. The polymer is expensive, since meta-distributed aromatics are, in general, dearer than para-distributed analogues. Nomex™ has similar properties to aliphatic nylon polyamides at room temperature but is considerably better at higher temperatures (up to 200 °C). It is difficult to melt-process and is generally provided in fibre form or as paper sheets by interfacial polymerization and extrusion into a non-solvent. A further example of the crankshaft effect is to use naphthalene derivatives instead of benzene ones. The fused second ring skews the polymer chain, but the monomers required are relatively costly.

The relative cost of starting materials has prompted investigation of mixed para- and meta-linked systems, where a modicum of the crankshaft effect is imported to enhance processability. A typical strategy for a polyester might be to employ quinol as a paradihydroxy compound, since the meta-diol is considerably more expensive than the meta diacid derivative, and to use a mixture of terephthalate and isophthalate monomers as the acid components. The diacyl chlorides could be used to improve reactivity, but it is less desirable for polyesters than polyamides to allow the generation of hydrogen chloride as a byproduct, so an alternative strategy exploits a transesterification reaction in the reverse sense to that previously encountered in Chapter 4. Instead of a volatile alcohol such as methanol being evolved to drive over the equilibrium, the alcohol component is used as the ester of a volatile acid. Quinol diacetate evolves acetic acid during the polymerization to give the appropriate molar mass.

Transesterification with commercial polymers

Since polyethylene terephthalate is a cheap and readily available material that contains terephthalate units (see page 90), another strategy towards chain rigidity involves transesterification of this polymer with an appropriately reactive but more expensive quinol derivative or polyester containing quinol units. This evolves ethylene glycol during the reaction and yields essentially a random copolymer with a high degree of aromaticity in the chain, punctuated by occasional flexible sp^3 carbon links from residual ethylene glycol units.

Substitution on the aromatic rings

A final strategy to disrupt the strict linearity of the aromatic linked high performance polymer chain is to put groups in the rings to induce steric interactions that hinder crystallinity. An obvious example is to use, for example, a methyl substituent (i.e. replace an aromatic proton by a substituent group R in Figs 6.3 or 6.4). However, whether as part of the diacid or on the diol unit, the starting materials are relatively expensive, and even when copolymerized with somewhat cheaper unsubstituted units the materials are less cost-effective.

Main chain liquid crystalline polymers

High performance polymers of the types described above are sometimes termed 'main chain liquid crystal polymers' because rigid materials can display certain co-operative properties associated with solids when they are either in the melt or sometimes in solution. Such high performance polymers as can be melted show thermotropic liquid crystallinity, while in solution Kevlar™ (which cannot be melt-processed) displays the less common lyotropic (solution) liquid crystallinity.

There are subtle consequences of this phenomenon, since the liquids show anisotropic shear dependence of viscosity. This means they flow more readily in the direction of shear than normal polymer solutions. Special processing apparatus is required and a company desiring to produce items made from these materials has an additional start-up cost. Furthermore, differential cooling of a melt near the edge of a mould, and the presence of necessary additives such as mould-release agents and lubricants, further complicates the moulding process and mitigates against simplistic economic considerations when addressing potential applications. Nevertheless, high performance polymers remain a useful example of structure–property relationships applied to polymeric materials.

Other functionalities are available, with consequence upon properties, and polyether etherketone (PEEK), (Fig. 6.5) is a successful commercial polymer with 'high-performance' mechanical properties.

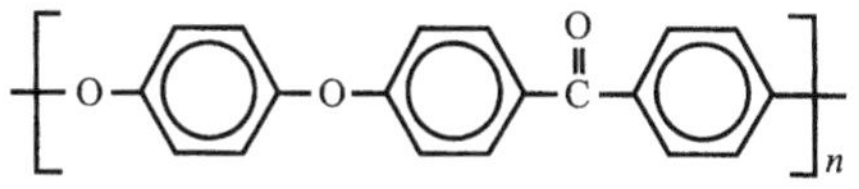

Fig. 6.5 Structure of polyether ether ketone (PEEK).

6.3   Electrically conducting polymers

Principles

Polymers are best known for their effectiveness as electrical insulators, and electrical wiring throughout the world is now sheathed in plastic. However, it was recognized some time ago that polymers with an appropriate structure ought to be able to conduct electricity, but the same features that might allow this phenomenon also introduce intractability and processing difficulties, and hence the science of conducting polymers had to wait until the mid 1970s when better defined materials started to be made. There are now numerous polymers with substantial conductivities, as can be seen compared with inorganic materials in Fig. 6.6. The backbone structures of the most important conducting polymers are given in Fig. 6.7. (Note—these are different systems to those obtained by adding conducting fillers such as carbon black to conventional polymers—see page 50)

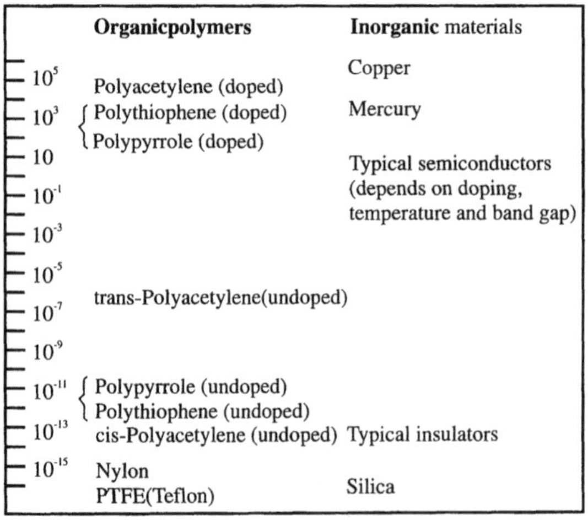

	Organicpolymers	**Inorganic** materials
10^5	Polyacetylene (doped)	Copper
10^3	Polythiophene (doped)	Mercury
	Polypyrrole (doped)	
10		Typical semiconductors
10^{-1}		(depends on doping,
		temperature and band gap)
10^{-3}		
10^{-5}		
10^{-7}	trans-Polyacetylene(undoped)	
10^{-9}		
10^{-11}	Polypyrrole (undoped)	
	Polythiophene (undoped)	
10^{-13}	cis-Polyacetylene (undoped)	Typical insulators
10^{-15}	Nylon	
	PTFE(Teflon)	Silica

Fig. 6.6 Typical conductivities in S/cm of various polymers and inorganic materials.

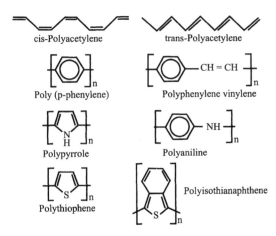

Fig. 6.7 Typical structures for conducting polymers (shown as uncharged forms).

For electrical conductivity it is necessary to transfer charge along a conjugated chain, between chains, and also along grain boundaries or between particles as shown schematically in Fig. 6.8. The most difficult process energetically will control the rate of charge transport and this will vary with the nature of the polymer, its physical form and other parameters, but in all cases conjugation along the chain is necessary although it is not sufficient for carbonaceous polymers simply to possess a conjugated chain. Figure 6.8 shows the polymer polyacetylene, which may be considered an archetypal system. Conducting polymers tend to fall into two distinct types, those similar to polyacetylene and those similar to polypyrrole, although with discoveries of yet more materials, the distinction becomes less evident.

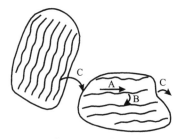

Fig. 6.8 Schematic of conduction pathway in a conducting polymer (a) intrachain; (b) interchain; (c) interparticle.

(NB: measurement of conductivity is not a simple task. Effects of AC and DC currents can be different and thin films may show different phenomena compared to, for example, compressed pellets or powders. Conductivities are usually quoted from measurement at low values of DC current performed by a method to avoid contact resistance phenomena.)

Polyacetylene and related materials

The simplest conjugated polymer chain is a polyacetylene chain, which may be represented in shorthand in Fig. 6.9.

Fig. 6.9 Resonance of a neutral polyacetylene chain.

Charge motion, shown by 'curly arrows', produces a dipolar form. Such charge separation is counter to coulombic forces and thus requires a significant input of energy. Consequently, 'neutral' polyacetylene does not conduct at room temperature.

Another way of putting this is that all bonding orbitals are filled and the gap between the highest occupied (HOMO) and lowest unoccupied (LUMO) orbitals cannot be bridged in this temperature regime. However, if the HOMO is incompletely filled then a different situation arises. Now, for example, a positive charge (missing electron) could move along the chain without the problem of charge separation, as shown in Fig. 6.10. Note that the 'herringbone' pattern of double bonds reverses at the charged discontinuity.

Fig. 6.10 Resonance of a single positive charge (polaron) along a polyacetylene chain.

Removal of an electron to leave the positive charge is an oxidation reaction which can be performed by a chemical oxidant or electrochemically (Fig. 6.11). In both cases the positive chain requires compensation by an anionic

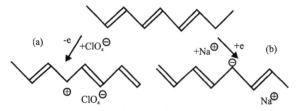

Fig. 6.11 Oxidation and reduction of polyacetylene to produce conducting forms. The insulating form is oxidized to produce (a) a perchlorate-doped material with a cationic chain; or reduced to produce (b) a sodium-doped material with an anionic chain. The conducting forms have typical conductivities in excess of 100 S/cm.

species from the system (chemical oxidants will have associated anions, while an electrochemical cell contains an electrolyte salt to carry current). This anion must penetrate the polymer matrix and diffusion also affects conductivity build-up. The depiction in Fig. 6.11 is simplistic, since many more than one charge will be present on any one chain. Unlike polypyrrole it is readily possible to make a polyacetylene chain negative by reduction, as also shown in Fig. 6.11.

So the production of conducting polyacetylene is a two-stage process. First the polymer is prepared in its neutral, non-conducting form. Acetylene (ethyne) and is readily polymerized by a Ziegler–Natta catalyst (Chapter 3) such as $AlEt_3/Ti(OBu)_4$ in toluene. Low temperatures favour the *cis* isomer while temperatures towards 100 °C favour the *trans* isomer. Although polyacetylene has this complication of *cis/trans* isomerization in the neutral form, the polymer automatically converts to the *trans* isomer in the conducting form and Fig. 6.12 shows the enhancement of conductivity as polyacetylene is exposed to iodine oxidant. Note the tremendous change as the same sample goes through an increase of 14 orders of magnitude in conductivity. Very few bulk materials can show a similar phenomenon and when it is also noted that a wide range of other properties such as density, porosity, surface energy, charge storage, colour and morphology also change with oxidation level, then the great present interest in conducting polymers is explained.

The charged conducting forms of conducting polymers are often termed in the literature 'doped', by analogy with semiconductors such as silicon, which becomes conducting upon 'doping' with small quantities of different valence species such as phosphorous atoms, although the charge levels are very different between semi conductors and conducting polymers.

Doping is an equilibrium and gaseous iodine is evolved from the doped polymer; thus iodine-doped polyacetylene is not useful in practical applications. However, this clean solvent-free system in which the solid polymer is doped with a gaseous oxidant remains useful for mechanistic and theoretical studies.

The most conductive polymer to date is polyacetylene prepared very carefully to avoid defects such as sp^3 carbons, cross-links or side products that interfere with conjugation. Other defects include carboxyl groups or hydroxyl groups from attack of oxygen, because a major drawback of polyacetylene is its sensitivity to air. The highest conductivity of doped polyacetylene is close to that of copper, but this does not mean that these materials will replace

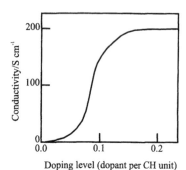

Fig. 6.12 Typical variation of conductivity with doping level for polyacetylene.

$$(CH)_x^{x+} . nClO_4^- \quad | \quad | \quad (CH)_x^{x-} . nLi^+ \; \rightleftharpoons \; (CH)_x^0 \, | \, nLiClO_4 \, | \, (CH)_x^0$$

charged form   discharged form

Fig. 6.13 Possible 'all-plastic' battery; electrode reactions exemplified by polyacetylene.

copper in wires. The ductility, pliability, strength and temperature-resistance of copper will always give it the advantage as a simple current-carrier, but the benefit of a conducting polymer is where other properties are required in addition to the ability to conduct electricity.

Such an application is one which exploits the ability of polyacetylene to become positively or negatively charged. This allows the possibility of an all-plastic battery as shown in Fig. 6.13. Lithium perchlorate is driven out of the electrodes as the cell is discharged, and then back in again upon recharging. Either half-cell could be employed with a more traditional battery electrode and many cationic-chain conducting polymers such as polypyrrole can be used as one electrode. At the moment problems of shelf-life, cycle life and stability mitigate against more widespread usage although there are promising niche applications.

There are other polymers similar to polyacetylene which are prepared in a neutral form before separate conversion to a charged conducting form. These include poly (p-phenylene) (PPP) and poly (phenylene vinylene (PPV) as shown in Fig. 6.7.

PPP requires forcing conditions (e.g. AsF_5 oxidant) to dope into its conducting form. This precludes routine use in conducting applications and it is actually commercially viable in its neutral form as an insulating coating.

PPV may be considered an alternating copolymer of polyacetylene and PPP. It is attracting attention because of its photophysical properties as the basis of a polymer light-emitting diode (LED) in which an electrical potential is converted into the emission of light. This requires the availability of charge carriers with appropriate energetics, but the polymer must be in the neutral uncharged form or it would short itself out rather than emit light. Normal LEDs are small-scale inorganic crystals such as gallium arsenide, and benefits of polymers include the possibility of large area displays, greater processability and the ability to tailor the emitted wavelengths of light to give a range of colours by control of structure.

PPV can be substituted on the aromatic ring (e.g. ortho–methoxy) and on the bridging unit (e.g. cyano replaces alkene H). This represents the sort of functional manipulation that is possible. Substituted monomers can be used as homopolymers or else copolymerized with other PPV units.

Precursor routes

The structural feature of extended conjugated double bonds confers rigidity in the polymer. Conducting polymers do not have carbonyl or amido functions in the chain as do the high performance polyesters and polyamides discussed in Section 6.2 so there is less hydrogen bonding, although there are polar interactions between the chains and the dopant ions in the ionomeric conducting forms. The result is that conducting polymers are not as strong as high performance polymers and can be more flexible. They do not dissolve as readily and attempts to melt them produce decomposition. Three main strategies have been employed to circumvent this problem of processability in the conducting form: two are to involve either employment of a functionalized monomer or to incorporate some other processability-enhancing components in the final polymer, and both strategies have been employed with the polypyrrole class of polymer and are further discussed below. The third strategy is used for the polyacetylene class and the idea is to make a processable precursor polymer. This non-conjugated, hence non-conductive polymer, tends to have more conventional polymer properties and thus can be processed into the desired format and converted to the conducting form afterwards. This is exactly the strategy employed for three-dimensional networks (Chapter 5) which have similar processing problems when in the final form and require the involvement of processable intermediates. Elegant precursor routes have been devised for polyacetylene, PPP and PPV. These are given in Fig. 6.14 and demonstrate the depth of subtlety in organic chemistry that can be employed in polymer science.

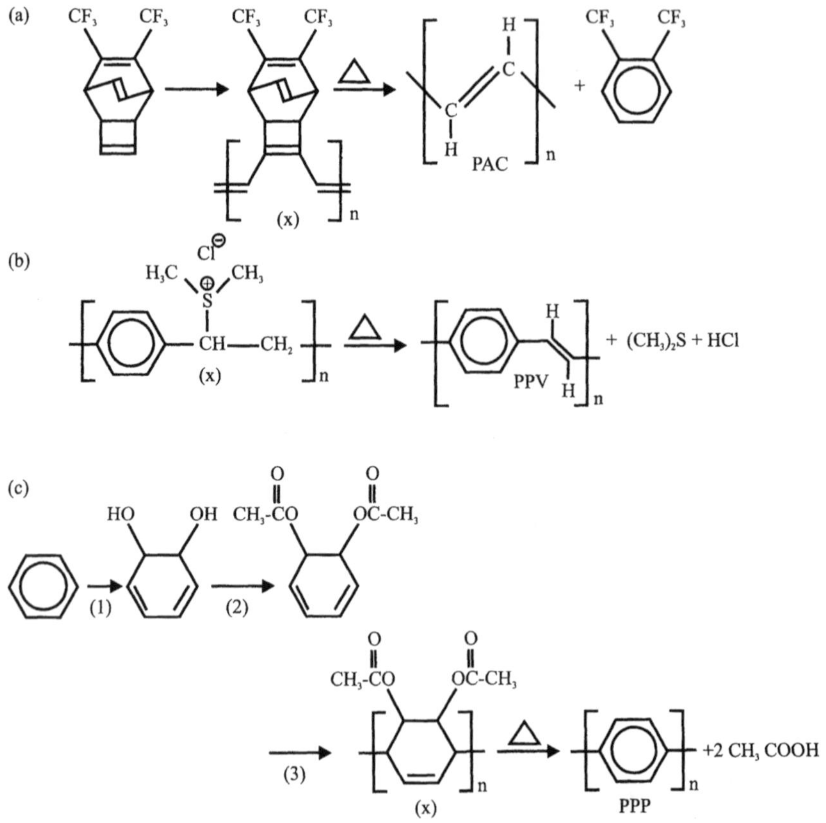

Fig. 6.14 Precursor routes to conducting polymers (a) polyacetylene (PAC), (x) is a soluble processable precursor; (b) polyphenylene vinylene (PPV), (x) is a soluble precursor polymer; (c) polyphenylene (PPP), (1) bacterial oxidation; (2) esterification; (3) polymerization to give the precursor polymer (x).

The route to polyacetylene was devised by scientists at Durham University in collaboration with British Petroleum and involves a ring-opening metathesis reaction (ROMP; see Chapter 3) to produce the processable precursor which then undergoes a thermal retrocycloaddition reaction to give polyacetylene with release of hexafluoroxylene. A number of variations on this route have been devised to control the final step in different thermal conditions with release of different molecules.

The route to PPV has been exploited by among others Cambridge University, the University of California and related commercial organizations and again uses an elimination reaction to produce the conjugated chain. Further sophistications have been introduced to produce derivatives of PPV, or else functionalized copolymers, and the scheme is representative. The route to PPP developed by ICI involves a neat bacterial oxidation as a key step.

Polypyrrole and related polymers

This is now by far the most widely studied class of conducting polymers. Pyrrole (or similar monomer) is oxidized either chemically or electrochemically to give the polymer directly in its positively charged conducting form, charge-compensated by anions from the oxidizing medium. A plausible mechanism is shown in Fig. 6.15. Oxidation is the loss of electrons, and here the monomer forms a radical cation that dimerises with proton loss.

The intermediate dimer has longer conjugation than the monomer and hence oxidizes onwards in the reaction conditions to give the polymer, which must be predominantly linked at ring positions next to the nitrogen atom to maintain conductivity. It is found that there is one charge per three or four pyrrole rings, across a range of preparation conditions. It is more difficult to obtain an intermediate doping level in this polymer than it is for the iodine doping of polyacetylene (Fig. 6.12). In the pyrrole structure the anion could be Cl^- and $FeCl_4^-$ from $FeCl_3$ or an anion from the electrolyte.

Polypyrrole requires a single step to be produced in the conducting form and, in the reverse sense to polyacetylene requires a reduction step to convert to the neutral form. Reduction gives a tremendous conductivity change accompanied by changes in many other properties including colour. A very thin film of polypyrrole is dark green–blue in the conducting form and yellow in the neutral. For polythiophene the respective colours are blue and orange, which offers greater promise for electrochromic (electrical colour

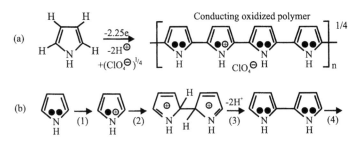

Fig. 6.15 Formation of a representative heteroaromatic polymer, exemplified by polypyrrole (a) overall process; (b) proposed mechanism, (1) monomer oxidation; (2) dimerization; (3) proton loss; (4) repetition.

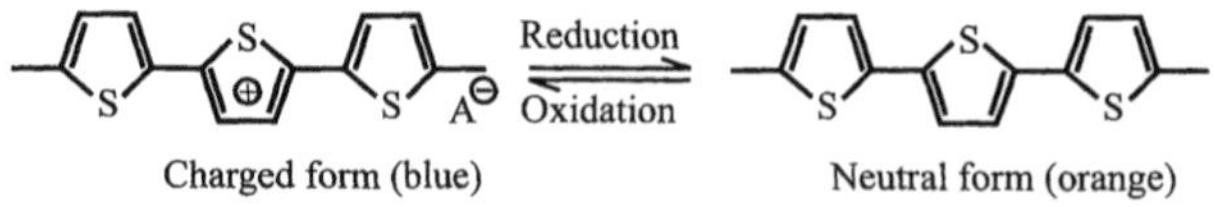

Fig. 6.16   Redox behaviour of heteroaromatic conducting polymers, exemplified by the electrochromism of polythiophene.

changing) displays as shown in Fig. 6.16. An important point is that these polymers offer opportunities for large area displays.

Other effects of redox are useful, particularly since these polymers seem intrinsically more biocompatible than metals. They can act as electrically switched perm-selective ion-exchange membranes and, if the simple anion BF_4^- is replaced by a bioactive anion such as salicylate (aspirin) or glutamate, then an electrically driven drug delivery system is possible to release specific amounts by external control of the reduction current. There is also a change in size and solvent-swelling properties during redox and thus a possible application is 'artificial muscles' in which electrically controlled expansion and contraction is exploited.

Polypyrrole is surprisingly tolerant of preparation conditions, particularly given the intermediacy of cationic species which in principle could be attacked by nucleophiles. It can even be prepared in water and is stable in both air and water in its conducting form (unlike polyacetylene). It is less stable, however, to oxygen in its neutral uncharged form although other polymers of this class, such as polythiopene, suffer less from this problem. Polypyrrole is the easiest conducting polymer to prepare. A very wide range of chemical oxidizing conditions will produce a powder, while electrochemistry in a wide range of conditions also produces freestanding films that can be peeled from an electrode if desired. Subtle differences between samples require accurate description of preparation conditions to maintain reproducibility. This is because conducting polymers in their ionomeric forms are non-stoichiometric compounds and thus are different to conventional polymer systems where properties and structures are less dependent on exact preparation protocol.

A further benefit of the polypyrrole class of polymer is the range of structural manipulations that can be performed (Fig. 6.17). Conductivity may be diminished to some extent compared to the simple polypyrrole archetype, but often other beneficial properties may be enhanced.

1. Vary the ring system X = N for pyrrole, S for thiophene
2. Vary substituents, keeping the X positions free for polymerization. (R_3 is present if X is nitrogen)

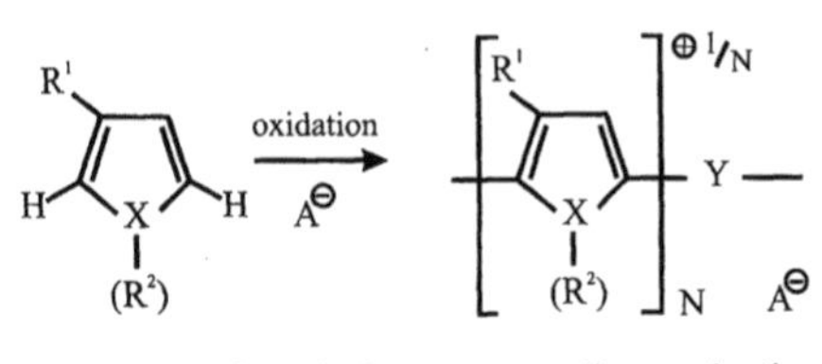

Fig. 6.17   Structural manipulations in heteroaromatic conducting polymers.

3. Nature of anion (A^-)
4. Doping level ($1/N$)
5. Copolymerization (Y is a different unit)
6. Post-treatment to alter structure.

All of these approaches can produce acceptable conducting polymers. If the substituent group R is alkyl or alkoxy then the polymer can have solubility in organic solvents while, if R is an alkanesulphonic acid chain or its salt (e.g. $-CH_2 CH_2 SO_3^-$), then the polymer has some solubility in water in its charged form because the anionic dopant is attached permanently to the polymer chain. These so-called 'self-doped' materials have yet further ion-exchange properties.

Quite complex groups such as dye molecules (photoactive), redox species (e.g. ferrocenyl groups) with further electron transfer effects, co-ordinating ligands (e.g. bipyridine units to sequester metals such as ruthenium), or electrocatalytic species can all be attached to the monomer ring. If these are so bulky that the pure monomer cannot polymerize then copolymerization with non-functionalized (and non-sterically hindered) pyrrole monomer allows formation of a copolymer in which the functional species are present proportionally along the chain (i.e. exemplifying the principle shown by the symbol Y in Fig. 6.17). Many other copolymer systems can be proposed, even with conventional polymers, with suitable chemistry. The nature of the dopant anion A^- also affects polymers more than might be expected. Even changes between simple organic and inorganic anions such as ClO_4^-, BF_4^-, $CH_3SO_3^-$, trifluoromethanesulphonate, p-toluene sulphonate and other sulphonate salts can significantly alter conductivity, elasticity, flexibility, permeability, morphology, and other physical and mechanical properties. However, more complex anions may be incorporated, such as sulphonated derivatives of phthalocyanines (photoactive and catalytic species), bipyridines (metal co-ordination complexes), sulphonated conventional polymers acting as polyanions (e.g. polystyrene sulphonate) and bioactive anion species such as heparin (the aim being to coat metal implants in medical systems to avoid rejection since heparin is the natural blood anticoagulant). More than one anion can be incorporated, such as a small amount of a functionalized anion bolstered by a simpler one.

Another consequence of biocompatibility is that nerve regrowth may be facilitated upon a conducting polymer template. This addresses a longstanding problem in medicine. It is also possible to entrap proteins and enzymes in a polypyrrole film as it is formed. Glucose oxidase continues to act when immobilized in polypyrrole and there is considerable scope for novel biosensors with this technology.

Composites

Polypyrrole has an ability to blend with conventional polymers giving a trade-off in properties between the two polymer systems. A charged conventional polymer (such as polystyrene sulphonate) can be used as polyanion dopant, but a simpler approach is to saturate a conventional polymer (chain- or step-growth depending on compatibility) with for example pyrrole vapour, and then expose to an oxidant, or else saturate the conventional polymer with an oxidant and then perfuse with the monomer. Electrochemical polymerization can be achieved by moving strips of conventional polymers perfused with the monomer over electrodes that are also rollers in a film-producing process. Laminates, latices, emulsions and suspensions may also be made. Applications such as anti-static (very important to protect computer chips) or shielding from electromagnetic interference (also important because of the density of emitters and receivers for communications in modern life) do not require great conductivity, so a balance between processability and conductivity may be achieved. There are a vast range of systems which may be considered and composites of many heteroaromatic conducting polymers are widely studied.

Polyaniline

Polyaniline (Fig. 6.7) is not strictly a heteroaromatic monomer but has a number of similarities to polypyrrole (chemical oxidation to 'aniline blacks' has been long known as 'pyrrole blacks').

Similar studies on electrochemical and chemical oxidation, redox behaviour, the presence of additional functionality, processability etc. have been performed as for polypyrrole and its derivatives and polyaniline has unique features. There is an extra equilibrium in the switching between conducting and insulating forms, as shown in Fig. 6.18. The figure also gives representative chain units in the various forms (3–6). (There are three redox levels, each of which may be protonated or deprotonated.) Of these, the pH-driven switch between emeraldine salt and emeraldine base is attracting interest (shown as structures 1 and 2 in Fig. 6.18).

Polyaniline is less tolerant of preparation conditions than polypyrrole and the list of anion dopants used in the preparation is more limited. However, subsequent replacement of preparation anion by a dodecyl benzene sulphonate makes polyaniline become soluble in solvents such as *N*-methyl pyrrolidone (NMP), or m-cresol, and can be spin-coated or otherwise solution-processed. This anion exchange is an example of post-treatment, which is another way to modify conducting polymer properties, although not widely used for polypyrrole.

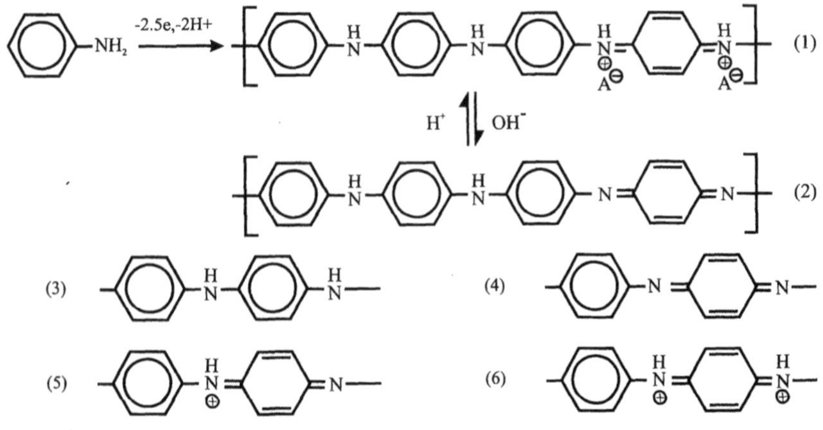

Fig. 6.18 Various forms of polyaniline, interconnected by redox and pH equilibria.

Functionalized anilines may also be polymerized , although the derivatives so far studied are less complex than those for polypyrrole.

Polyaniline-composite textiles have also been prepared and processing features of the polymer are useful. In addition, in Victorian times the capabilities of 'aniline black' (deep green when dispersed) to protect against corrosion were appreciated and iron seats at railway stations, exposed to the weather, were coated in a polyaniline-containing lacquer.

In summary, conducting polymers provide a classic example of structure–property relationships in which the inherent macromolecular nature of a polymer combines with functionality to produce features that lead to applications well outside the normal run for conventional polymers.

6.4   Polymers with functionalized side chains

Principles

This provides a large class of useful polymers, and the full range of polymer-supported reagents provides a breadth beyond the scope of this *Primer*. The idea is to employ traditional polymeric physical properties from the backbone chain while adding external properties from the side chains. Ion-exchange resins are a simple example. A key feature of any exchange chromatography system is that a reactive stationary phase be employed that is readily separated from the eluent containing the desired target molecule. Polymer beads can be made accurately to desired dimensions and, if suitably cross-linked, will not dissolve in the eluting solvent. Polystyrene cross-linked with divinyl benzene is a popular system. If the beads are derivatized such that the surfaces contain exposed acidic groups (e.g. sulphonic acids, phosphoric acids, carboxylates, etc.) or basic groups (various amine functionalities) (Fig. 6.19), then passage of a pH-sensitive species past the beads will result in ion-exchange equilibria being set up. Optically active immobilized species can also enhance chiral separation. The beads can contain other functionalities to entrap reaction intermediates or products. The important feature is the ease of separation of the polymer-bound reagent from other species, and the desired product may either be deliberately attached to the polymer, and so removed with the polymer to a separate vessel and recovered, or else left in the original solution when unwanted species which attach themselves to the polymer are removed.

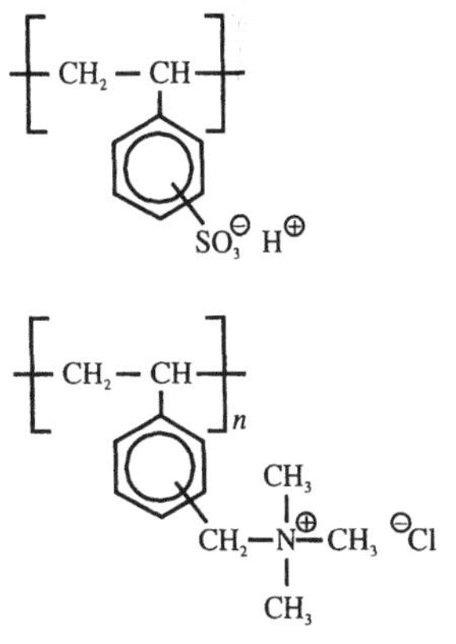

A sophisticated example of the principle involves the polymer coating of magnetic iron oxide to give 'magnetic beads'. If these are then also coated with, for example a biological antibody or similar species, then this can 'clamp on' to its target biomolecule from a mixture. The beads can then be removed by a magnet for separation and onward treatment.

The list of chemical reagents that can be attached to polymers by various strategies is long and there are important modern developments based in particular upon the ease of removing a polymer-based reagent from a reaction mixture. The first of these was the Merrifield synthesis of amino acids, joined in the correct order by peptide (amide) linkages.

First, a polystyrene bead is converted to the chloromethylated derivative by electrophilic attack and then the desired amino acid previously protected at its amino group, displaces the chlorine by nucleophilic substitution using its

Fig. 6.19 Examples of functionalities used in polystyrene ion-exchange resins.

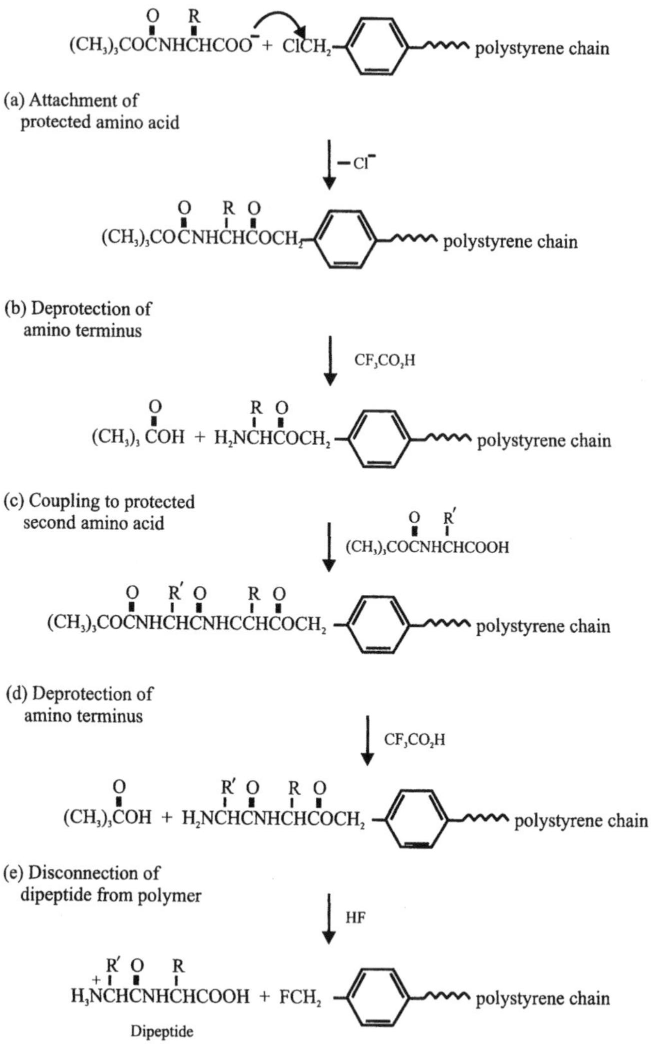

$(CH_3)_3COCNHCHCOO^- + ClCH_2$—⬡—polystyrene chain

(a) Attachment of
protected amino acid

↓ $-Cl^-$

$(CH_3)_3COCNHCHCOCH_2$—⬡—polystyrene chain

(b) Deprotection of
amino terminus

↓ CF_3CO_2H

$(CH_3)_3COH + H_2NCHCOCH_2$—⬡—polystyrene chain

(c) Coupling to protected
second amino acid

↓ $(CH_3)_3COCNHCHCOOH$

$(CH_3)_3COCNHCHCNHCCHCOCH_2$—⬡—polystyrene chain

(d) Deprotection of
amino terminus

↓ CF_3CO_2H

$(CH_3)_3COH + H_2NCHCNHCHCOCH_2$—⬡—polystyrene chain

(e) Disconnection of
dipeptide from polymer

↓ HF

$H_3^+NCHCNHCHCOOH + FCH_2$—⬡—polystyrene chain

Dipeptide

Fig. 6.20 Merrifield synthesis of a dipeptide on chloromethylated polystyrene support.

carboxylate group. The protecting group is removed and now another protected amino acid is added via a coupling agent. Further deprotection and reaction with yet another protected amino acid allows the desired peptide to grow away from the polystyrene bead. Final scission with hydrogen fluoride removes the peptide from the bead as shown in Fig. 6.20.

The whole process can be automated, since the beads are readily separated for filtering and washing after each step, whereas side-products remain in solution for disposal. The synthesis, named after the 1984 Nobel Prize Winner, would be very cumbersome if all steps took place in solution, with traditional isolation and extraction of products after each step.

The principle has now been extended in the development of combinatorial chemical synthesis (sometimes called 'chemical libraries'), as shown in

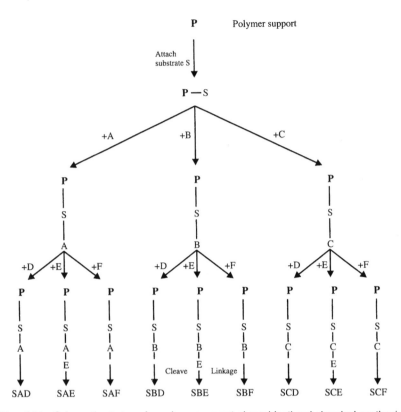

Fig. 6.21 Schematic strategy for polymer-supported combinational chemical synthesis (chemical library).

Fig. 6.21. This provides a rapid screening system, for example for the preparetion of a number of members of a class of compounds which might have pharmaceutical activity. An initial substrate molecule is attached to a polymer before the sample is divided and exposed to different reagents A, B and C. Each product is further divided and exposed to reagents D, E and F. The benefit again is that the products remain attached to the polymer throughout the manipulation, until the initial linkage (which must be labile in appropriate conditions) is cleaved to liberate each of the nine possible products (3 × 3) as shown. Tens of thousands of products can be obtained very rapidly by this procedure which has had a considerable impact upon synthetic chemistry.

Side chain liquid crystal polymers ('comb' polymers)

Here a useful functionality is attached as a side chain to a polymer which is not in the form of a solid bead. The materials are instructive as to the typical strategies available for permanent attachment of side chains to conventional polymer backbones.

Thermotropic Liquid crystals show co-operative behaviour over the temperature range just after they start to melt. Essentially, to exhibit this phenomenon, a molecule needs to have a high aspect ratio, i.e. its length should be several times its cross-sectional area. Examples are given in Fig. 6.22. Such rod-like

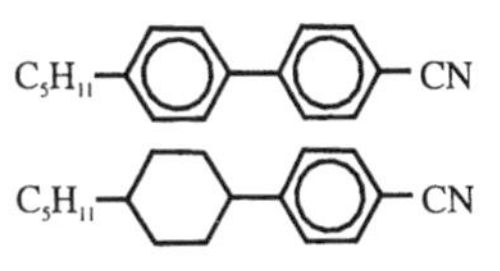

Fig. 6.22 Representative small molecule liquid crystal compounds.

molecules cannot easily tumble past each other when lattice bonds are broken at the onset of melting. Instead, a number of ordered liquid forms can occur as planes of molecules start to float apart, still keeping interplanar integrity, leading to a smectic (soap-like) phase. The liquid looks milky due to scattering by particles of similar dimensions to the wavelengths of visible light. Further heating produces disruption of intraplanar forces, but still the molecules cannot freely tumble and hence the nematic (thread-like) phase is formed. Finally, further heating provides sufficient energy for tumbling to give the optically transparent normal isotropic liquid with random orientation of each molecule. If the molecule has a dipolar structure the polar vector remains in the ordered liquid forms to interact with electromagnetic radiation and allow the molecules to respond to an electric field. The liquid crystal displays, for example in wristwatches, employ a polarizing filter to provide a directional sense to light passing through the liquid crystal phase, the ordering of which can be modulated by the electric field produced by the watch battery. Photoactive molecules may also exhibit non-linear optical effects (for example, frequency doubling) which are of interest in the development of optical data storage, 'optical computers' and other optical equivalents of electronic systems.

The importance for polymers is that when a small molecule liquid crystal is attached to a suitable polymer backbone chain by a sufficiently flexible spacer linkage, and, if there is not too high a density of such side chains (i.e. not on every monomer unit), then co-operative liquid crystalline behaviour and associated phenomena from the side chains will still be observed. Side chain liquid crystal polymers offer flexibility, robustness and strength over small molecule analogues, and in display systems suffer less from parasitic degradative electrochemical discharge at the electrode (which is meant to be there only to induce an electric field and not to initiate electrochemistry).

The attachment of liquid crystal side chains is typical of strategies for side chain polymer modification and extension to other chemistries for attachment of other functionalities can easily be imagined.

The commonest backbones used are polyacrylates and polysiloxanes. Each requires a different end-group on the spacer chain for attachment. Figure 6.23 shows an alcohol-ended unit that can be attached to, for example, polyethylacrylate (polypropenoic acid ethyl ester) by a simple transesterification reaction, yielding an equilibrium polymer containing azo side units and unreacted ethyl groups.

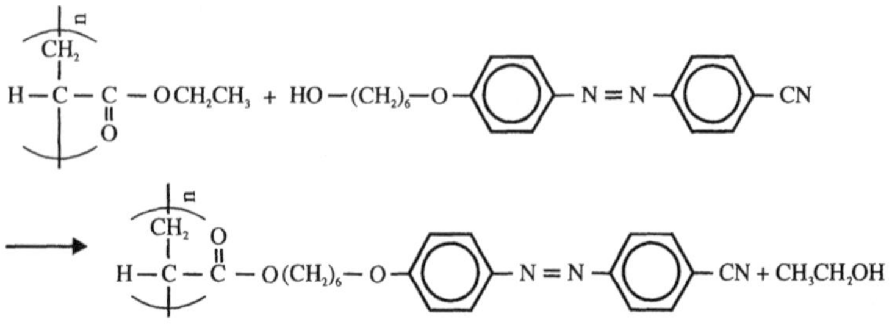

Fig. 6.23 Attachment of hydroxy-ended liquid crystal side chain to poly(ethylacrylate) by transesterification.

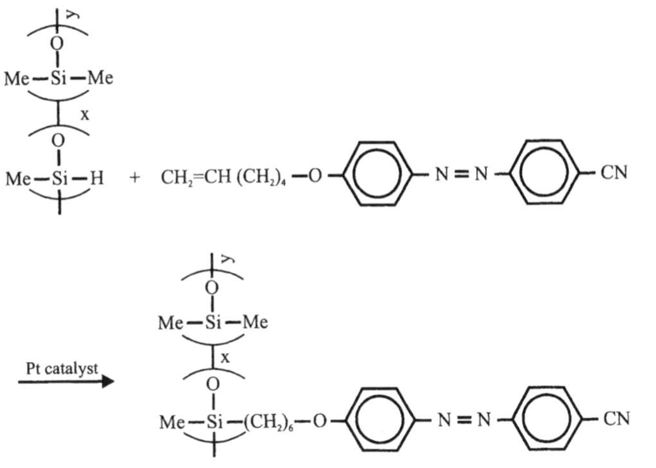

Fig. 6.24 Attachment of alkene-ended liquid crystal side chain to a polysiloxane.

Figure 6.24 shows the chemistry of attachment to a polysiloxane, and employs the reactivity of the Si–H bond (see Chapter 5) which adds to an alkene-ended side chain in the presence of tin or platinum-containing catalysts.

6.5 Polymeric photoresists

In the production of printed circuit boards for microelectronics, or in compact disc manufacture there may be over 20 different layers of materials. Laying down of these with appropriate microscopic definition often involves protecting one area of the surface to let another process occur only on desired regions. This is achieved by polymers that either react to cross-link under irradiation, or else decompose under irradiation. The former protect a defined area against reaction in the next step while the latter degrade the polymer coating to allow subsequent reaction on the previously covered area.

If acrylamide (CH_2=$CHCONH_2$), a divinyl cross-linker, and a photo-sensitizer are irradiated at an appropriate wavelength (often in the ultraviolet region), then the exposed areas become coated in an insoluble polymer, while washing with water removes unreacted monomers from areas that were masked from light. Alternatively, organic-soluble polyvinyl cinnamate (an analogue of polyvinyl acetate (Chapter 3) with PhCH=CH COO replacing CH_3 COO in the pendant group) may be spin-coated on to a surface and, where irradiated, becomes cross-linked by cyclodimerization so again cannot be washed off. Similarly, heat-treated (and precyclized) polyisoprene may also cross-link under irradiation in the presence of a suitable sensitizer.

In the reverse sense polysilanes (which contain direct Si–Si bonds and are more reactive than siloxanes) degrade upon exposure to light such that the polymer is lost from the exposed area, but remains intact under the mask. Other systems include novolak phenol–formaldehyde resins (Chapter 5) with naphthalene diazoquinone sulphonate (NDS) as photosensitizer. Photolysis causes complex chemistry with evolution of nitrogen and rearrangement to form a more soluble polymer that is washed from the exposed regions. Considerable developmental effort has gone into the optimization of the specialized polymer systems.

6.6   Polymers in the environment

The functionality in this case is the role of the polymer in the ecosystem. This is a large topic and this *Primer* is closed with a few comments on key issues. These include bioproduction of useful technological polymers, remediation and biodegradability, and waste treatment and recycling. All of these have long-term implications for the ecosphere.

Many polymers, particularly polyalkenes (polyethene, polypropene), polyhalocarbons (PVC, PTFE) and aromatic-containing step-growth systems (PET, polycarbonates, some polyamides) are very resistant to microbial degradation. In ultraviolet-stabilized grades with antioxidants they are difficult to break down by exposure to the elements and so remain in the environment for a long time. They also require consumption of non-replaceable petroleum for their feedstocks.

They can be destroyed by incineration and so return some useful heat energy, but can cause emission into the atmosphere of catalyst residues and species derived from additives, which may involve toxic heavy metals and other problematic materials. In principle, polymers based on carbon, oxygen and nitrogen could be burnt cleanly, but halogen-containing polymers have difficulty with the fate of the halogen atoms. PVC must be incinerated at very high temperatures otherwise partial combustion leads to cyclization reactions with the formation of toxic dioxins, while PTFE evolves fluoro species.

The key aspect is to know which polymers are present in the waste sample. Only when one is sure of polymer identity can recycling become feasible.

PET can be reused directly as a lower grade material following reprocessing; for example, it can be ground up and reconstituted as insulation boards to replace polystyrene, where the higher melting point of PET *versus* polystyrene (260 °C *versus* 100 °C) allows use of hot-melt adhesives in board preparation.

Step-growth polymers can generally be hydrolysed back to their original components for re-use, but chain polymers require thermal degradation back to the monomer, usually in forcing conditions. At present this is costly, but the economics may become overturned if environmental legislation forces the issue. The disposal of car tyres is an obvious chain polymer target to be addressed.

Another strategy is to develop polymers which are more readily degraded in the environment. However, there must be a balance between loss of desirable properties and decomposition rate. For engineering applications attack by environmental vectors could be dangerous, but is acceptable for high volume low grade packaging materials.

When polyethene is prepared by free radical reaction in the presence of carbon monoxide a small number of carbonyl-derived units appear in the chain. Polyethene is usually quite resistant to photodegradation, but this makes it open to attack by sunlight when inappropriately discarded in the environment.

Decomposition by bacteria and micro-organisms is attractive for polymers left in landfill sites, but many materials are quite resistant: for example, PVC, PET, polypropene and polystyrene seem to be immune to enzymatic attack. Natural biopolymers offer opportunities for remediation. The difficulty in practical applications is finding materials with the correct physical properties and processing characteristics. One approach is to blend a bioactive species with a traditional polymer. Thus plastic bags can be made from polyethene mixed with the cheap oligosaccharide (sugar-derivative) starch. The idea is that attack on the starch by micro-organisms in a landfill will break the bag down to smaller particles.

Some synthetic polymers are more readily attacked by biovectors than others. Thus polyurethanes, particularly polyether–polyurethanes, can be degraded over time and their general biocompatibility allows use in disposable artificial sutures and in other biomedical applications requiring self-destruction.

There are many esterases (enzymes that hydrolyse esters) in the environment but they do not attack aromatic esters such as PET. Aliphatic esters are open to bioattack but generally have less desirable physical properties than aromatic ones (see Chapter 4), though they are suitable for some applications. The synthetic polymer polycaprolactone (Fig. 6.25a) is susceptible to environmental degradation, as are hydroxy-functionalized polymers such as polyvinyl alcohol and its ester derivatives (e.g. polyvinyl acetate, see Chapter 3) and cellulose derivatives provided that not all the hydroxy groups are functionalized i.e. there is a free –OH available.

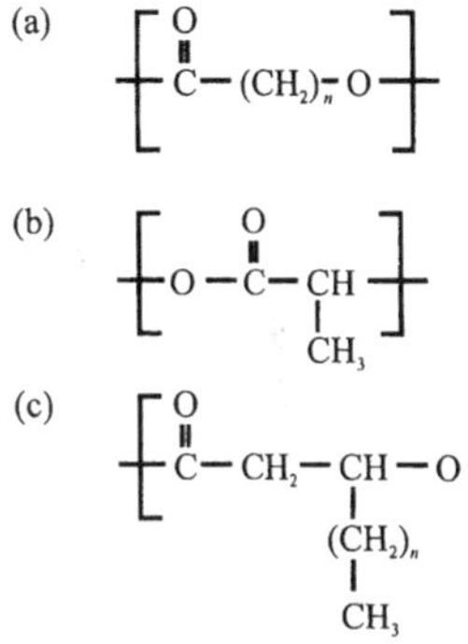

Finally, a most desirable situation is one in which a relatively simple and processable polymer can be produced naturally on a sufficient scale for practical usage, and also be biodegradable. Cellulose fits this description but requires the processing strategies given in Chapter 1. Another class of natural material, first identified in 1925 but now better appreciated, is aliphatic polyesters, exemplified by polyhydroxy butyrate (Fig. 6.25c). This is synthesized as an energy storage medium by the bacterium *Alcaligenes eutrophus* and is obtained in 100% isotactic stereochemistry from bacterial cultures. Some mechanical properties are similar to polypropene (flexural modulus, tensile strength), but with poorer extension-to-break. It is useful as a packaging material and is completely biodegradable. Polyhydroxyvalerate (Fig. 6.25c) is a homologue, also produced by bacteria. Polylactic acid (PLA) is a semi-synthetic material produced by step-growth polymerization of lactic acid, which is obtained by fermentation. In this case the product of polymer degradation can be metabolized and so PLA offers promise as an encapsulant for controlled release of pharmaceuticals and agrochemicals such as fertilizers and pesticides. It can also be used for sutures and other medical implants that are required to dissolve.

In summary, polymers have moved on considerably from the simple systems used only for their mechanical properties. This chapter has tried to give a flavour of the type of consideration important in modern functionalized systems.

Fig. 6.25 Repeat units of biodegradable aliphatic polyesters (a) synthetic ($n = 5$ for polycaprolactone); (b) semi-synthetic polylactic acid; (c) natural ($n = 0$ for polyhydroxybutyrate and $n = 1$ for polyhydroxyvalerate.

7 Conclusions and further reading

The aim of this Primer is to give a flavour of the breadth of chemistry covered by polymeric materials and to explore the sophistication behind these deceptively simple-seeming structures. In addition, polymers provide a convenient opportunity to study industrial aspects of chemical science. In fact, considerations of economics, availability of feedstocks, use of specialized catalytic reactions and similar matters are found throughout the chemical industry, but polymers provide particularly instructive examples. It is an appropriate time to review polymers since there are many new aspects and applications in prospect.

This is a very wide subject to cover in a single short Primer and readers are encouraged to investigate more specialized bibliography for areas of interest.

There are numerous reference books on all aspects of polymer science and technology, including series of volumes extensively filled with data, and over the years many excellent textbooks have been produced on polymers. The volume by Saunders "Organic Polymer Chemistry" is particularly recommended and a number of tables in this Primer have been adapted from this work.

The following list is intended simply to be illustrative.

Bibliography

Introductory texts

Nicholson, J. W. (1991). *The Chemistry of Polymers*. Royal Society of Chemistry, Cambridge.

Campbell, I. M. (1994). *Introductory Synthetic Polymers*. Oxford Science Publications, Oxford.

Hall, C. (1989). *Polymer Materials* (2nd edn). Macmillan, London.

Fried, J. R. (1995). *Polymer Science and Technology*. Prentice Hall, New Jersey, USA

Intermediate texts

Saunders, K. J. (1994). *Organic Polymer Chemistry* (2nd edn). Chapman and Hall, London.

Allcock, H. W. and Lampe, F. W. (1990). *Contemporary Polymer Chemistry* Prentice Hall, New Jersey, USA.

Munk, P. (1989). *An Introduction to Macromolecular Science*. John Wiley & Sons, New York.

Elias, H.-G. (1993). *An Introduction to Plastics*. Verlagchemie (VCH), Weinheim.

Young, R. J. and Lovell, P. A. (1991). *Introduction to Polymers*. (2nd edn). Chapman and Hall, London.

Specialized texts

Brown, R. P. and Read, B. E. (ed.) (1984). *Measurement Techniques for Polymeric Solids*. Elsevier, London.

Albertsson, A.-C. and Huang, S. J. (ed.) (1995). *Degradable Polymers, Recycling and Plastics Waste Management*. Marcel Dekker, New York.

Clough, R. L., Billingham, N. C. and Gillen, K. T. (ed.) (1996). *Polymer Durability, Degradation, Stabilisation and Lifetime Prediction*. American Chemical Society, Advances in Chemistry Series no. 249.

Ed. H. Singh Nalwa (1998) 'Handbook of Conductive Organic Molecules and Polymers' Volumes 1–4, John Wiley.

Acknowledgements

The authors thank all at Coventry University for their support, particularly Jenny McGlone for her help with a large part of the manuscript and Clive Dixon and the Teaching Resources Department for most of the figures.

D. J. W. also wishes to thank A. Jenkins and N. Billingham for exciting his original interest in the subject, all at Raychem Swindon (1981–1986) for rekindling it, W. MacDonald, D. M. Smith, D. Pullen, L. S. A. Smith, C. E. Hall, A. Chyla, M. Morris, D. R. Rosseinsky, R. Lines, J. E. McIntyre, and numerous others for continuing to keep the interest fresh and invigorated at various times.

J. P. L. wishes to thank D. C. Pepper and the late Eric Catterall for introducing him to the subject and to the many others in particular Dave Forrest, Dave Kershaw, Kate Fiddy, Dirk Kirschneck and Liz Brookfield who provided the stimulus for a better understanding of the discipline.

Last, but not least, both authors wish to thank their families for their continued support during the preparation of this *Primer*.

Coventry 1999

Index